The Contemporary Animator

The Contemporary Animator

John Halas

Focal Press
London and Boston

Focal Press
is an imprint of Butterworth-Heinemann

First published 1990

British Library Cataloguing in Publication Data

Halas, John
The contemporary animator.
1. Cinematography. Animation
I. Title
778.5347

ISBN 0–240–51280–4

Library of Congress Cataloging-in-Publication Data

Halas, John.
The contemporary animator / John Halas.
p. cm.
Includes bibliographical references and index.
ISBN 0–240–51280–4
1. Animation (Cinematography) I. Title.
TR897.5.H32 1990
778.5′347—dc20 90-13825

Phototypeset by Scribe Design, Gillingham, Kent
Printed in Great Britain by BPCC Wheatons Ltd, Exeter

Contents

Introduction

Today, most of the world's population absorbs its information, receives its entertainment and learns its lessons in the form of moving pictures. More and more time is spent watching television, video cassettes and films than ever before. Apart from sleeping and working, viewing has suddenly become the most time-consuming human occupation in our civilization. A major sporting event like the Olympiad is watched by a billion people. At the same time, a popular, animated feature-length movie is being seen by several million viewers year after year. Moving pictures, especially animation, which is considered to be the most international medium, are gradually taking over from the printed word. For centuries, printed texts prevailed, dominating the mental outlook of previous generations. Today, they have been replaced by the hypnotic effects of moving pictures, whose influence can be compared with that of Byzantine and Renaissance art hundreds of years ago.

There are certain fundamental factors attached to moving compared with static pictures, no matter how they are made – the capacity of moving pictures to engage the public on several levels (sometimes physically, sometimes emotionally, sometimes intellectually) and their ability to convey a range of effects which other media would find much more difficult to achieve. The synthesis of pictures, movement, sound and effects, and the newly available computer manipulation of three-dimensional images, are means of communicating with audiences and of expressing a deep feeling which have not existed until now.

From a visualizer's point of view, engaged in animation, this is a unique chance to reach audiences in millions for the first time. However, it would be misleading to assume that to achieve a level of proficiency, let alone outstanding performance, is easy. The process of production today is far more complex than before, as is the great variety of new equipment, styles and materials at one's disposal. The range of options is enormous. A starter faces the dilemma of what particular aspect of this industry to specialize in, what equipment to choose from the many available, how to define priorities, and what particular media would suit his or her talent.

It is interesting to note that animation has only become respectable during the last ten years. It is now almost accepted on the same level as other media, craft and art practices. The number of students entering colleges has doubled during the last decade, and the proportion attending the full length of the course is higher than in the other study categories. In spite of the fact that animation needs a great deal of patience, hard work, concentration, industry and precision, interest in the medium is constantly on the increase. Why?

First, it combines a variety of activities to appeal to the contemporary creative mind: visual invention in two and three dimensions, composing or assembling music and sound effects, writing dialogue and commentary, harnessing electronics and optics, and maintaining organizational discipline over all these studies. Second, it is one of the most direct means of achieving self-expression, especially if one is in control of the production process and if the end product corresponds to one's original idea. Third, it can be profitable. If the work is in advertising it is likely to be well paid, so is the field of special effects, where there is an acute shortage of skilled artists and expert craftsmen.

Apart from talent, which is the first essential, one needs practice and experience. It is a field where one should give enough time to become technically knowledgeable. Compared with other visual communications media (ballet, painting, photography, graphics, live-action films), animation is expensive in labour and material. At college level, as a rule, it is taught with limited resources. The problems which constantly arise are how to make the best of such resources, how to plan one's time, and how to choose techniques, style and equipment accordingly.

The attention focused on animation, its importance in society, the new opportunities in the field, the expansion of technical facilities and the availability of different electronic systems means that a total revaluation should take place in thinking about animation, since the old concept cannot do justice to it. Therefore most of the present forms of tuition require updating to meet the requirements of contemporary animation.

1 At the beginning

If you want to be an animator try to get into one of the colleges specializing in the genre. There are many in the USA, Canada, the UK and Europe. All have adequate facilities to practise with 16 mm optical cameras, some with computer graphics equipment.

Basic animation skills

Among the other visual communication disciplines, such as graphic design, fine arts, illustration, film and photography, I maintain that animation is the most difficult. It contains aspects of most of the other disciplines to various degrees, plus others such as a sense of time and space, and an understanding of music and motion. Animation is a rich medium, demanding a particular devotion. Students are expected to be prolific, with an enquiring mind, patience and diligence. Heavy demands are placed on one's intellectual capacity and one's technical capabilities. What are the basic requirements for a student animation project?

- To invent an idea which can only be carried out in animation
- To visualize such an idea in terms of animated film, highlighting the major points in the idea in time, continuity and sound
- To have the essential skills to make the idea work in motion, and to be able to design the objects, forms and characters to complement it
- To design the backgrounds, establishing the visual style of the film
- To load a camera, photograph scenes of the film with the right lighting, unload the film correctly in order to have it developed by a laboratory
- To manipulate individual drawings through an animatic machine and test movements
- To assemble the individual scences and edit them in continuity
- To record sound effects, dialogue commentary and music
- To mix the separate sound tracks and dub them onto a single strip of magnetic track
- To evaluate the end result objectively

For many of the above activities there may be specialist help available at the college to assist the student; also it must be decided whether the whole process is to be an individual or a group activity. The list omits several fundamental steps for the production of an animated film, since what has been described is only workshop practice to provide students with a technical grounding.

In an evaluation the first questions to consider are, has the exercise achieved its objective? Would a different stylistic approach have been better? Has the film taken too long to make? While, in most cases, it is a good idea to begin a student with workshop activities, at a later stage many theoretical studies should be added, especially in view of the growth of electronics.

It is interesting to note that since the introduction of tuition some years ago at design colleges, many student films have won major international awards. The latest is the National Film School's film *The Hill Farm*, made by Mark Baker, which won the Grand Prix at Annecy International Animation Festival in 1989. The success of student films can be attributed, apart from their genuine talent, to the circumstances and conditions of work. These are total concentration on the task without interference by clients, a reasonable timetable, allowing work to progress according to its natural pace, and relative freedom from budget pressures. It is valuable for a student to spend from two to four years on studies comparatively free of commercial pressures, since such conditions will not occur in the commercial field. For most students, therefore, this may be the most creative period of their lives (which is, as a rule, difficult for them to realize).

Tuition methods

Naturally, the nature of tuition can differ significantly from one country to another. Animation schools in the USSR are similar to the USA's Disney crash course, both motivated by the decline of skills in fluid animation techniques, since the old generation of director/animators is gradually passing away, and the classical art of fluidity in animation is no longer in fashion. Feature-length films have created a new market for character animators, and once again there is a demand for that unique skill. Especially in Soyuzmultfilm in Moscow, experienced animators spend a high proportion of their time in teaching the younger generation, since it is realized that their type of animation is primarily children's entertainment, which depends on competent fluid animation techniques.

A notable departure from routine tuition is at Los Angeles in the Animation Department of CalArts, where Jules Engel, an outstanding artist in his own right, teaches his pupils in a more fundamental way. Visual appreciation receives high priority, as does graphic design skill. 'Thinking before doing' is adopted before animation is studied. A wide range of stylistic approaches follows and animation's rich potential is proved. As to a formal curriculum, few animation courses take much notice of this. The six- to twelve-period tuition is primarily based on practical workshop activity. Few understand that, in animation, the not so obvious can matter much more. To make an idea visible, to express emotion in a dramatic format, to convey humour and comedy in visual form are highly tangible matters, but as Jules Engel proves, a good way to teach is to train the mind first.

There are other forms of teaching animation, and a number of schools offer animation courses at a very early age, such as that of Kati Macskassy in Hungary. These courses combine storytelling capabilities with drawing practice and, for sound, music skill and voice exercises. Children's natural instinct for creation is fulfilled by such tuition, which is very popular in the art periods at schools, stimulated by television.

Most of the colleges base their systems on past technologies and present practices. Few include computer media, which everyone knows is the next phase of development in animation. Although tuition in depth may be obtained in computer graphics at the Middlesex Polytechnic in London, unfortunately the same facilities do not exist elsewhere. The industry is desperately short of computer graphics artists, which is a rapidly growing activity, providing a fast and versatile approach to presentation and the final production. It is in this direction that student knowledge should be extended, and where the future generation of animators should experiment with new ways to express themselves and search for a new animation language.

Production studios

No strict formula exists for what part of the production a studio should do. Every one is different. It depends on the country and the personalities working in the unit. Nevertheless, to avoid large expenditure for overheads and regular staff most studios work with as small a team as possible, hiring out their own specialized services and subcontracting others as suits the situation. There are now many freelance animators doing their work in their homes and producers conducting their business from telephone booths. This is also a time of opportunity for specialist animators partnering their wives, husbands or friends. Major clients, advertising agencies or TV campaigns prefer small teams. They are often cheaper, fitting in with budgets, and can be more creative. Such clients know that small units have to be supplied with

Twenty stages of producing an animated film:

1 Treatment
2 Storyboard
3 Timing
4 Recordings
5 Character design
6 Layout
7 Animation
8 Assistant animation
9 Line test
10 Background
11 Tracing
12 Painting
13 Checking
14 Photography
15 Viewing
16 Editing
17 Pre-recording
18 Dubbing
19 Negative cutting
20 Answer print

storyboards and soundtracks. They do not expect the studio to supply anything beyond the rush print.

Under continuous work pressure, a small team has a habit of becoming a medium-sized one. The larger it becomes, the more of the 20 stages of production it can undertake. The comprehensive service, which includes storyboarding, writing and recording, may start with a permanent staff of 20 in traditional cel animation, or 12 if the studio is based on computer graphics. For a television series with constant output the studio requires an integrated staff of 40 upwards, a point where industrial output may force the management to employ production assistance and as many internal departments as it can afford. In the case of large and very large studios such services are essential if they are to achieve uninterrupted, smooth production. In the case of specialized studios the supply of stages narrows down to three or four if there is only painting, tracing and checking to carry out. Frustrations due to mistakes which can occur force clients to add photography as a fourth on the list, and today most studios shoot the painted cels as part of the service they provide.

A very substantial evolution has taken place in the production of animated films in the last generation. Today, the small studios dominate the field, with three to five individuals each carrying out a wider range of functions and being flexible in their work. Few large studios, with a hundred personnel and more, remain on a permanent basis. Those that are left usually work continuously on feature films and television series for a home market which is able to support them; these are in Japan, the USSR and Hungary. In the USA studios tend to group together until a project is completed, disband afterwards and regroup for another project. The labour-consuming aspect of the work is usually subcontracted in areas where labour is cheap and plentiful, the Far East being one obvious choice.

Styles, skill, experience, timetables, equipment, space available and the size of the budget will obviously determine whether a project is to be made in a small, medium or large studio. For practical purposes the composition of the unit may be as follows.

The individual animator

Such a person prefers to work alone, and may be a genius pursuing an idea which can only be carried through by that special pictorial touch that none can copy. Such a project can take years to make, requiring a lot of resources, time and money. The contribution of such individuals can be substantial on an aesthetic level but requires continuous financing. The other type is the freelance animator working at home. These are professional animators whose contribution is highly sought after, and remuneration is in terms of footage produced. As a rule, they are highly expensive, but the output contains professional skills of a specialized nature. An example is Borge Ring, working in Holland and servicing feature-length animated films produced in Denmark, Germany and England. The number of two-person partnerships has increased substantially. Possibly the best known is the Italian team of Gianini and Luzzati, whose work is based on individual and complementary skills, with Luzzati contributing the visual graphic aspect and Gianini the technical execution to both the story outlines and the filmic concept. One may say that they are a perfect combination. However, no matter how significant the individual contribution of an outstanding animation artist, most animated films are made by large groups. The following are some possible structures according to the flow of work and the type to be carried out.

Units of three to six personnel

Taking the Bob Godfrey Studio in London as an example, one may observe the logic of such a limited structure. Bob Godfrey acts almost as a one-man band in his series *Henry's Cat*, doing the storyboard, timing it, drawing the key positions for animation, design and backgrounds, editing, and even making the soundtrack. The stories are supplied by a freelance writer, Stan Hayward. Since there is not too much follow-through activity, two assistants are sufficient to complete and photograph one 5-minute episode in one week. In this way, it is possible to supply high-quality series at a reasonable price that satisfies everyone. Even a small unit requires a production assistant, a fact which is often ignored by those freshly starting. Purchasing, invoices, orders, payments, typing, recording and storing records of scene movements have to be dealt with on a practical level if a smooth flow of work is to be achieved. As a rule, the difference between small and larger units lies in the level of delegation to others and not in the work undertaken, which may be very similar.

Units of seven to 12 personnel

Most work is being made in studios of this size. It is ideal for maintaining a reasonable flow of commissioned work for advertising agencies, television companies requiring animation inserts and industrial organizations requiring informational diagrams. A unit even of this size needs an expert producer who knows the medium of animation sufficiently to be able to negotiate costs, timetables, styles and techniques with the client. Such a producer may be the head of the unit and may also act as a creative director. One or two animators could be attached permanently to the team with an animation assistant and a small team of tracers and painters on constant call, avoiding hold-ups in emergencies.

Commissions do require commitment of time and reliability of deliveries, especially if television air dates are involved. For a team of this size it is safer to run a camera service in-house. Until teams evolve their own computer graphics and work entirely with magnetic video systems, the optical film process will continue to require laboratory services for printing and developing their film rushes. Sufficient time, often in short supply in a small unit, must be added to their delivery dates for these functions.

Units of 13 to 20 personnel

Hibbert Ralph Animation in London is an example of such a unit. The studio services a broad range of markets, from TV commercials to children's entertainment and animation for TV openings. The personnel are able to carry out several functions, according to the type of work in hand. The unit is capable of expanding quickly, and since it is planning a television series it is likely to double in size in the near future. Like most other studios, when the workload justifies it a number of freelance specialists are hired on a temporary basis.

Units of 40 to 60 personnel

Kecskemet Studio in Kecskemet, Hungary, a well-integrated studio some 80 kilometres south of Budapest, has been in continuous production since 1970, with its own camera set-up for optical 16 mm and 35 mm. It has a permanent staff of six animation directors with assistants, layout and background artists and inking and painting departments. The studio has been specially built and designed for the production of animation. Consequently it is comfortable and functional, and generates a professional atmosphere. Its versatility in style is wide, according to the nature of the subject, from children's series to experimental shorts and commissions from outside the country as required. Its output, depending on the complexity of the assignment, ranges from 10 to 20 hours of animated films a year, which makes it one of the largest studios in Central Europe.

Units of over 100 personnel

The growth in the market for animation has split large studios into three distinct categories:

- Units which are brought together for a specific objective to produce a single feature-length film, such as those produced by Ralph Bakshi (*Fritz the Cat* and *Heavy Traffic*) and TVC London (*Yellow Submarine* and *When the Wind Blows*);
- Permanent units like Soyuzmultfilm in Moscow, with a personnel of approximately 500 in constant production for 50 years;

- Service studios such as those in Taiwan, the Philippines and South Korea, concentrating on tracing and painting and providing labour at a competitive price for producers in the USA and other Western countries.

The most difficult situation is the temporary unit brought together to carry out a single assignment.

The temporary studio

The function of this type of studio resembles that of a large musical orchestra brought together for one performance. It requires more preparation, and more time for getting to know each others' weaknesses and strengths and for adjustments to the principal characters, styles of design and the type of animation most suitable for the film, all of which require patience and strong management. The twenty successive stages of production must be preplanned almost like a military campaign, controlled and checked for quality and volume of output, and coordinated on an economic level. Since the major part of the work is still hand-made, one is constantly exposed to unforeseen human problems, which is an unmeasurable factor in animation and often dominates a feature-length production.

As a rule, the production personnel become very important in such a unit, essentially concentrating on the progress of production and the coordination of details.

The film should be divided into sequences of 3 to 6 minutes in length, and each sequence may have an animation director. If this is not possible, for example when producing an 80-minute feature, each animation director could be responsible for several such sequences. The supervising director has the task of coordinating overall aspects of the feature with the producer a step further, and is not involved with physical execution. When the director and producer are the same person the problem is simplified. In the anatomy of a unit, not sufficiently appreciated (but essential) are the animation and colour checkers, who make certain that no mistakes are let through the production, since once a scene is completed it is very expensive to repeat it.

The objective is to achieve a creative flow of work, a task not easy to achieve when the work is basically repetitive. It is for this reason that it is advantageous that such parts of the production as tracing and painting should be done by mechanical means.

Permanent units of over 100 personnel

Large studios like Soyuzmultfilm must have a firm foundation both in the marketplace and in economics. Since this studio is state-supported in any case, it is secure and able to concentrate on production. This does not mean that their operation is not cost conscious. There is control of finance and time in the studio and an internal check on the comparative expenditure of each department, as with other types of film production. The basic character of a large studio lies in how the different types of activities are broken down. At that scale, the studio's activities become an industrial process: the task of management is to maximize its physical output without too much interference with artistic content. The balance between quality and quantity is constantly under survey and becomes more delicate when there is pressure for more and more products and demands by individual artists for more time for their work. Such opposite interests, of course, apply to all large-scale studios, no matter where they are. In Moscow the work is divided (according to the capabilities of the animation directors) into smaller units of activity, with each unit being allocated whatever technical and manpower facilities are required. There are also central facility services such as final photography, laboratory and sound dubbing.

The output of the studio is, by Western standards, low but constant, considering that the technical equipment is installed to assist manual labour. However, the studio manages to supply the market with a great number of children's television series, occasional feature films and entertainment shorts. Among these is the work of such outstanding artists as Eduard Nazarov and Feodor Khitruk, and a number of promising newcomers. There is a tolerant attitude towards such stars as Yuri Norstein, considered to be one of the best animation artists of our time, who runs his own unit and can take several years over his films. For instance, Gogol's *The Overcoat*, his latest film, has been in production for three years, with 12 minutes completed in two years

Lately, apart from competition by the Kiev, Tallin, Alma Ata, Riga and Tbilisi Studios, all of whom work independently from Moscow, the local television station has established its own production unit which will supplement the supply of productions substantially. Since the TV studio also supplies titles, openings and animation inserts, its services are technically more advanced, with the latest versions of computer graphics equipment. The structure of the individual units is well balanced with assistants and clean-up animators, layout and background artists, and follow-up services if the film requires celluloid rendering. However, during the last few years there has been a move towards using cutouts and clay animation, which is in line with what is happening in the rest of the world. Most of the individual units which make up the studio are headed by experienced animators, advancing from lower grades and learning their craft through practice, which makes the studio a conglomerate unit and one which recognizes individual capabilities. Production management has the intelligence to recognize the fact that even a large studio requires individual talent to survive.

Lately, all large studios have added a separate department for the exploitation of merchandise linked to their characters and stories in the form of toys and books. There is a strong interrelationship between production and merchandise, since, as a rule, the latter has been proved to be more successful commercially than the former, especially in the West.

Large studios in the West are different. It is a paradox that, with the exception of Hanna and Barbera, both of whom were originally individual animation directors, many of the others were headed by business managers not entirely conversant with the medium. Filmation, for instance, once a studio with over 300 personnel and the largest in Hollywood, after several takeovers by TV corporations ended up with business supervisors. The unit did not last too long after that. The same occurred with the Marvel unit, which, during the period of its existence, numbered well over a hundred personnel. One may come to the conclusion that large units are not entirely conducive to individual creative output, even when the end product is a mass-produced article such as a children's television series, for daily transmission.

2 Production methods

Two vital factors must be given consideration in every production: idea content and visual presentation.

Idea content

All manufactured articles, from mousetraps to aircraft, must start with an idea. In the case of an animated movie (essentially a manufactured article, no matter how individually it is made), the process from idea to finished product has become more and more complex. Painters and photographers are able to convert their ideas much more directly into a finished article. Their success or failure depends only on their own creative and technical ability. In the case of an animator, especially with a flow of continuous production, the case is very different. His or her ideas must filter through many minds and many pairs of hands. The end result, therefore, does not depend entirely on the animator's skill alone. The idea, passing through many stages, must be constantly kept alive. One often finds that in the process of delegation an idea tends to lose its freshness or originality during each stage of production.

After fifty years of the adoption of the conveyor belt system, which was invented to achieve higher speeds and greater output, there has been a return to the personalized working method, with one person in control of all the stages of production and execution of details. The expansion of animation colleges during the last 20 years has also contributed to this system substantially. The development of computer animation is following the individual approach concept as, apart from a programmer, the turnkey system is no longer required; the animator can carry out all the stages on the computer.

The storyboard

Once a satisfying idea has emerged there is only one way to control it. Write it down in the form of a short synopsis, describing how the idea will be treated visually, and then start drawing it. The idea will grow in the form of its visual presentation. The 'storyboard', as it is generally known, should eventually become the project's bible. It should be given as much consideration as possible, and a lot more than it receives generally.

To start with, this is the stage where the suitability of a project is determined. It is also where the animator's ability to bring to bear his or her own personality and feeling for the audience emerges.

The major consideration should always be the central concept. Out of this grow the supporting ideas to give shape to the work. The storyboard is an essential part of any motion picture. For an animated film it is indispensable. A verbal description of an idea may be acceptable for a theatre play but for animation it is the visualization of the idea out of which flow the inventions which could develop, perhaps, into an entertaining, lyrical visual poem or a dynamic, dramatic film. If the idea is for an experimental film, the maker is the final judge, unless the production takes place in a college, where the tutor gives the final approval. If the work is sponsored it is essential to obtain the client's approval. Several stages of discussion can take place until the storyboard is found suitable for the client's requirements. For this reason, it is better to start with a set of thumb-nail sketches and evolve the ideas as more finished drawings are made.

In the case of advertising films the process may be different. Most advertising storyboards are designed by the advertising agencies themselves, and are then developed up to the stage of an animated presentation for the client. The presentation also contains voices and sound effects tracks which approximate the finished product on the screen. Often even the timing is fixed, leaving little room for the personal contribution of the animator, who, when chosen to finish the commercial, provides a specialized animation service. Since such work is very well paid there seems to be no shortage of units to carry out the task. Such units are, as a rule, chosen according to their technical facilities, graphic design, animation and creative capabilities, and sympathy with the ideas implied in the storyboard. However, it is seldom that the characters, choreography or motion continuity can be changed at this stage. Consequently, if the agency's storyboard designer does not understand animation and fails to realize the talents of the animator, the end product could fall short of expectations and even the best idea can suffer.

The storyboard should be an exploratory platform in any animated film. Essential aspects of it are flexibility and fluidity. It should reflect what the animator has in mind. The graphic quality may not matter much at this stage, provided that the ideas are conveyed to others who will be involved in carrying them out. But even here the choreography should be clearly implied, the continuity set and the action emphasized. Like a human torso, it should be well balanced, the episodes should relate to each other and the climaxes should be in the right place. When such a stage is reached it should not be too difficult to put the flesh on, to build the project further towards its ultimate destination: the completed film or tape.

As a general guide, depending on the style, length and type of finished work, it is customary to produce 30 to 40 sketches for each minute of action. The quicker the action, the more drawings are required to pinpoint the key moments of movement and any specific changes which will be required in the animation to carry out the idea.

For a feature-length film the storyboard, which will involve a number of animation units, must contain as much detailed information as possible. Once a long film is in production it is unlikely that there will be time to return to the basic concept of the story. Such a storyboard could take a very long time to emerge, but while it is in the process of creation it could serve as a centre point for discussion and for any changes in balancing the necessary dramatic development and highlights in the story. While such a storyboard may not contain the same qualities as a finished film, since it consists merely of rough sketches, today it is possible to video line test it, and add a voice track. Such an 'animatic' tape shot instantly could be useful on many levels: presentation to a client for approval, carrying out speedy alterations, and showing it to all who are involved in the production. In the case of factual exposition one need not go into such details, nor when the production is the work of one individual who will personally carry through all steps of production.

The usefulness of a functional storyboard has been recognized as a vital step in the process of production. An animated film, unlike a live-action documentary, cannot be shot off the cuff. Every step must be preplanned. Nothing can be left to chance. Nevertheless the storyboard is only a chrysalis, which is better left discarded once the heated battles and discussions about the film-structure are over and the real work has begun. It is at this stage that one should determine what particular technique one should employ to carry out the production.

Visual presentation

Visual styles determine what an idea will look like on the screen and what technique can do justice to it. Most beginners start with a particular style but find that this has to be adapted during production according to the technical limitations of the work. In the end they will emerge with a range of styles adapted to the technique chosen. It matters a great deal whether one works on one's own, with a small team or with a large one, and what kind of technical facilities are available. We have seen that there are different types of production organization: the unit for a large output mainly for mass-produced television series and feature films, with a minimum of 50 personnel; the small service unit with six to 15 personnel servicing commercial commissions; and the individual unit with one animator working on experimental productions alone or with a few helpers. The newcomer is the computer animator, either with his or her personal equipment at home or as a part of a video studio's service unit with an integrated workstation. The different approaches require different techniques, mentality and philosophy. With adaptation of any one of these systems, or a combination of them, one could cover all the techniques known today, from the simplest to the fully animated lip-sync type, from the strictly technical and factual approach to the highly creative and imaginative project.

Choosing the right technique is as vital as having good idea. One might assume that a well-conceived storyboard with its defined content and characters would indicate the technique best suited to the visual presentation. This assumption would have been true a few years ago, but it is no longer so. There used to be only two options: should the animation be flat, or should it be puppet? Today there are at least thirty options, and this often causes a major dilemma for film makers. The rapidly expanding computerized graphic systems are having a significant influence on the animation industry.

Expert computer animators – such as Judson Rosebush in New York, Art Durinski at Toyo Links Corporation in Tokyo, and Tony Pritchett in London – all maintain that computer animation is not at all like normal animation. Each of them sees the contrast from a different perspective. Rosebush maintains that it is mathematics that sharply divides the two. Durinski thinks that it is a totally different visual language, one which does not mix with the other. Pritchett maintains that since computer graphics is born out of science it will never complement hand animation, which is based on artistic emotion and expression. It is possible that they all are right. But there is no reason why each method should not enrich the other, and each benefit the other's practices. However, animators have so far been unable to adapt to the touch-button technique and scientists, who invented digital visual systems, have not given sufficient attention to animation.

In spite of this unresolved situation, full consideration should be given to the present state of machine intelligence. If computer animation is to be used it is imperative to know its capabilities.

The storyboard must be prepared in terms of such capabilities in an accurate form with mathematical precision.

This is contrary to the flexible nature of a storyboard prepared for hand-made animation where, as far as style is concerned, changes can be made. The users of computer techniques are full of warnings. First: if your idea is beyond the capability of computer graphic systems do not touch it. Second: if you can carry out a scene better manually, do not use the computer. Contrary to initial expectations of cutting costs by replacing expensive hand labour, computer graphic systems are still very expensive tools which can be highly dangerous in the hands of an amateur from a cost point of view. They demand a totally different relationship between the artist and the pencil, a different approach with special disciplines and intelligence. They depend on understanding how to apply numbers and codes, data structures, preferably in two- and three-dimensional space, building graphics based on points, lines, planes, pixels, volumes and voxels, different applications of the mechanics of digitizing, analog and digital conversions and the techniques of aliasing. All are vital elements for understanding how the new tools work and the effects one can achieve.

For many newcomers it is well worth acquiring the techniques. According to Jim Lindner, the New York computer animation producer, it takes two to three years after passing an art degree at a design college. But according to Richard Williams it takes ten years to make a professional animator.

The chart on page 16 outlines the main stages of production for hand-made animation and animation processed by computer. Whereas in hand animation there is not too much variety, computer animation is flexible, depending on the types of hardware and software being used. The first stages of production from idea to sound recording are the same in both systems, although one must be aware from the very beginning of the type of technique one will use, since this decision will influence all subsequent stages of the film.

Constant factors in animation

The changes which have taken place in the past few years belong primarily to the area of presentation, updating it using contemporary techniques. The fundamental forces which create the laws of movement – gravity and friction – cannot be changed. The Newtonian laws of motion have always been important in animation. They remain as starting points to be exaggerated, distorted, even defied. The forces of nature – wind, waves, tides, temperatures – are conditions of which one should also be aware. Natural motion is better left to the live-action camera, however, and should be avoided in animation. The art of animation starts

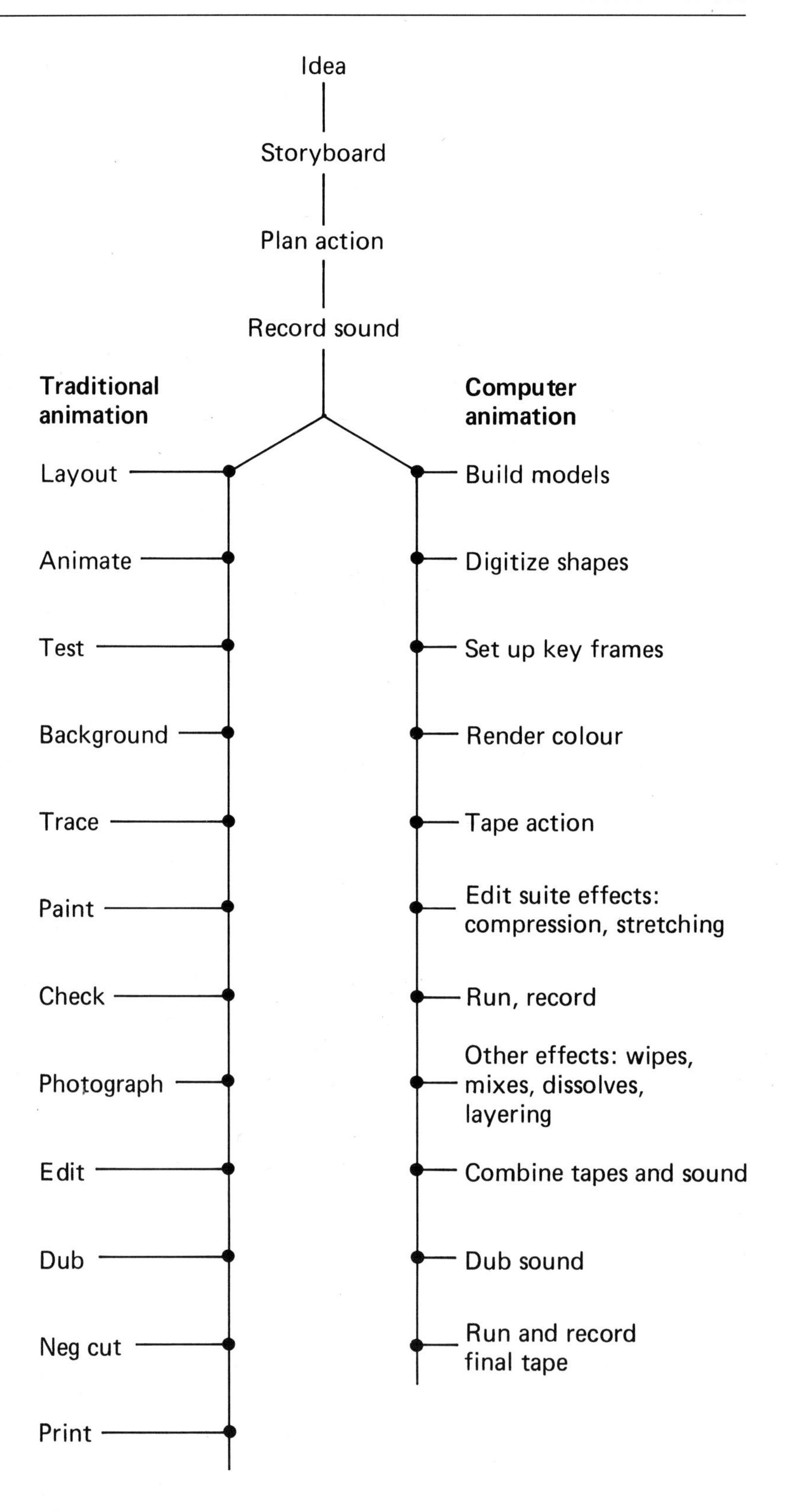
Idea
Storyboard
Plan action
Record sound
Traditional animation
Computer animation
Layout
Animate
Test
Background
Trace
Paint
Check
Photograph
Edit
Dub
Neg cut
Print
Build models
Digitize shapes
Set up key frames
Render colour
Tape action
Edit suite effects: compression, stretching
Run, record
Other effects: wipes, mixes, dissolves, layering
Combine tapes and sound
Dub sound
Run and record final tape

at the point where the live-action camera stops, a world of imagination only animation can create.

Another element which technical development does not change is the art of timing an action. The main influence here, which always has been a prime consideration in creating motion, is the overall visual intelligence of the general audience. People growing up in a visually complex environment, watching television from early childhood, can comprehend action far quicker than previous generations. This fact is exploited in TV commercials, where a message must be conveyed in a few (very expensive) seconds. Timing became a form of shorthand in visual communication, especially in animated films using simplified figures and objects, which are easier to identify as symbols. The basic principle therefore is unchanged. Quick movement requires fewer frames to draw, slow movement more frames. The positioning of objects and their size play a part, as do the characters' behaviour, mood and interrelationships, which influence their speed of action on the screen. Great liberties can be taken in animation. No comedy can succeed without exaggeration of time, especially the amusing effect of speedy action. For instructional purposes slow action may be essential to pinpoint certain facts (for instance, in mathematics and physics, when a principle must be proved and understood). Yet the opposite is the case in, for example, geology, when the passing of millions of years can be shown in a few seconds of screen time.

One learns by experience that realistic time photographed by a live-action camera looks unnatural in animation. Animation time has its own rhythm, its own structural principles. Each movement demands a different approach. Each character has its own individual style of walking, running and behaving – all the factors which make it a character. A sense of time is the main factor, irrespective of how a film is made.

Since half of animation is still carried out with traditional techniques, it is valuable to analyse them and describe the technical changes which have been taking place during recent years. The techniques practically all workshops practise are as follows:

- Animating flat (manual activities): cel, paper, cutouts, collage, engraving on materials.
- The range of three-dimensional approaches: puppets, wood, clay, plastics.

Celluloid

So far, more than 70 years since its invention, no better material has been found to reproduce sequential movement than the transparent cellulose acetate, commonly called *cel*. Why? It is easy to handle. It is flexible. It can be drawn on with ink, cel paint, cel pencils. Lines and shapes can be duplicated on it mechanically, and textural shapes and forms can be painted on it. But its basic value lies in economy. Four cels can be combined in each scene, and each level given a logical function. One cel can be held stationary (the body) while another closely matched to it can be moved (the limbs, say). A face can be held static while a mouth moves, speaking a line of dialogue. The combination of several characters is possible provided the cel levels are not varied. Otherwise tones and shades will flicker and spoil the effect on the screen.

It must be realized that the use of several overlaid cels slightly reduces the clarity of the picture. In the case of four cels (the maximum number advisable, depending on cel quality and thickness) the reduction in clarity may be 8–12%. This can be compensated for by the cameraman by adjustment of focus. Skilful manipulation of cel levels can determine how much labour is saved in the scene. An experienced animator can minimize the work by a clever combination of cels.

From Walt Disney's *Snow White and the Seven Dwarfs* (1938) to Frédéric Back's *The Man Who Planted Trees* (1987), most classic animation has been produced with the use of cel. For complex creation of movement it is unlikely to go out of fashion, not even with the newer technologies.

Cutouts and collage

To move a piece of paper around under a camera and alter its size or shape, moving one frame at a time, is an inexpensive way to create motion. Most beginners adopt this method to gain experience in animating inanimate objects. Much can be learned from it, provided something valuable can be developed later.

At the other end of the scale, artists of the calibre of Yuri Norstein (Soyuzmultfilm, Moscow) have proved that highly artistic and complex animation can be achieved with cutouts and collage provided one has the technical skill with the talent and ideas to back it up. Norstein justifies his use of the technique on account of the freedom it provides:

> My technique combines good and bad sides of animation which can be reflected in various ways. In my case it allows me to subordinate the entire artistic background created for the needs of a film to my inner ideas, to my artistic credo. On the whole, animated films are all the better for technical limitations. In an animated film everything must be created from the very foundation. One has to choose, and find all the necessary elements.

Norstein's contribution to the collage technique consists of using the simplest material with which he can express the most complex emotions in the most direct form. His characters are paper cutout figures joined with wire thread which allows each section of the body to be move frame by frame. They are developed with a sense of humour, emphasizing their nature and character almost as an extension of his own. His backgrounds, which are rich in tone and design values, are placed on a plane about 15 centimetres (6 inches) away from the plane on which he animates his characters. The camera is mounted directly above these planes and positioned so that both objects are in focus and the figures do not cast any shadows. In this way Norstein is free to apply textures in the design of his characters, making them richer in tone than most other animated ones.

The distance between the figures and the background also gives a sense of depth, which makes the scene appear to be half-way between cartoon and puppet animation. Norstein has a special aptitude for using this technique:

> An idea can be formulated in words, a means of expression in the visual arts not indispensable here. But here may be something more, which I may find difficult to put into words and which perhaps is not necessary for me. I have in mind ideas composed of concrete pictures, of the artistic context in which these pictures exist, of colours, tones, some kinds of flora, words, music, all these devices which can create something that cannot be translated into a language of words, that cannot be formulated as a definite idea of a film.

Norstein is considered as a contemporary poet who has found a new way of conveying his ideas. In fact, he has achieved more than that. He has bridged the biological gap between humans and animals by making his animals think and behave like humans. With the rich visual talent in his background painting he has also created an imaginary world, entirely original, that looks like no scenery we have seen in any other medium. All this has been achieved with modest pieces of paper, paint, scissors, wire and a single stop-motion camera. His team comprises one assistant and one camera operator.

Yuri Norstein is not the only one to have achieved so much with so little. Lynn Smith, while working at the National Film Board of Canada, has also shown what simple animated cutout figures could achieve in her film *This is Your Museum Speaking* (1979). Inanimate figures come to life and move with a convincing fluidity not expected in paper cutouts nor with such a rich rendering of shades.

Evelyn Lambert, formerly assistant to Norman McLaren, also from the National Film Board of Canada, has developed an excellent skill for the manipulation of cutout shapes in an abstract style, retaining the textural values of these shapes, which is one of the main advantages of the cutout technique.

However, the most natural rendering of cutouts comes from the Shanghai Studios in China. A large proportion of their work contains collage animation with animals and birds which lend themselves to this approach. The delicate forms and textures of the characters suit Chinese craftsmanship and sense of design. The story content is mainly for children's entertainment and complements the technique.

At a higher level the work of the veteran animator Te Wei reaches a perfection similar to Norstein's but in a different style. His film *Feeling from Mountain and Water* (1988) brings together the values of cutout shapes and the sensitive impression of elements such as wind, rain and water, and his sense of movement and filmic timing. It is also exciting and dramatic. The limitations and simplicity of the technique are again turned to advantage in the hands of a creative artist. The Grand Prix at the 1988 Shanghai Film festival was justly awarded to this film.

The degree of perfection achieved by some artists may take decades of practice, trial and error, and constant contact with the materials at their disposal. They have learned how to achieve a three-dimensional illusion out of what is basically a two-dimensional technique. This method is particularly suitable for beginners, as it is primarily a one-person operation or, at best, a small team one, not demanding a great number of specialists or large capital expenditure on equipment.

Painting pictures under the stop-motion camera

Since the stop-motion camera is fitted with a motor which automatically arrests the camera every single frame to take a picture, it is only logical to take advantage of this and draw or paint pictures progressively and expose them under the camera. We know that a rapid succession of individual images, when projected onto a screen, will appear to move in continuity. From a beginner's point of view, this is most likely to be the first step in finding out about animation. For an experienced visualizer it gives an opportunity to maintain his or her individual style. Take, for instance, the texturized motion paintings of Witold Giersz, the Polish painter, who advances the movement of his figures on his canvas frame by frame under the camera. He retains the characteristic brush strokes of a painted surface as the figures surge forward. His galloping horses seem as if the paint pours down on their tails and disappears suddenly when the brush dries up. However, such a synthesized motion is not so easy to achieve as it might appear. The input requires a sense of time, a preconceived choreography and infinite patience.

Another way of using such a technique is to create texturized figures on an unusual surface such as clear glass, as Caroline Leaf did in her film *The Street* (1977) at the National Film Board of Canada. The animation and textures of characters are achieved by the artist having total control over both. This time, the materials used are more sophisticated, ink and acrylic paints on a surface of sheet glass, the density of the paint being controlled with a sensitive touch of the fingers. This method is not recommended to be systemized but is used only as an example of how one may retain the total value of a moving shape and what can be achieved with inexpensive materials when one knows how to use them.

Not all materials lend themselves with such ease as inks and paints. Raoul Servais used etching to advance his movements, a most difficult and time-consuming task, in his film *Operation X70* (1971). The value of his film lies in the fact that only an artist could try almost any material to create the illusion of movement and to find new textural surfaces and ways of film making, no matter how difficult it may be.

Puppets

There are many new ways in which one can explore the advantages that three-dimensional plastic animation can offer. Once a puppet project is decided on, the first consideration is the material to be chosen. Next are the lighting problems in a three-dimensional situation, and third is the motion control of objects and characters. As in most other forms of animation, these are closely interrelated and all must be solved during the pre-production period rather than in the middle of a production.

Various materials have been used, from hard wood to flexible wires and plastics to soft clay, which today is the most popular material (until the next phase). Each material has a particular characteristic which requires a different feel by the maker before success can be achieved. They have one thing in common: the entire concept is three dimensional and the materials are manipulated in space.

The difficulties arise early in three-dimensional animation, and are mostly connected with the physical handling of the material. Here are a few. Clay materials can dry quickly; one may find that they have shrivelled or cracked during the night. They have a habit of showing fingerprints, which, in close-up, are magnified. If the lighting is too close or too hot they can melt, which distorts the characters. One may not have the essential sense of time and form for the continuous advancement frame by frame which is necessary in stop-motion with a malleable material, creating the wrong emphasis and not the effect which one may have planned. The popularity of clay (and today there is a wide range of textures from soft to hard) is due to the fact that one may develop the closest contact with it through one's fingers to such an extent that it behaves as an extension of one's body, and responds to the slightest command. If this skill is achieved and one can instrinctively control the timing and flow, one can maximize the unexpected expressiveness which only such a malleable material can provide.

Correctly manipulating clay may be the most direct way of creating emotions through expressions, movement, gestures and mood. It may also be the cheapest method of producing something from practically nothing other than a talent in stop-motion photography.

Wood is more difficult. It is a stiff material, offering the greatest resistance to being moulded and deformed. It is these characteristics which make it functional when creating characters out of it. Wood can be joined together in small sections by wires or rubber, plastic strips and ball joints, and each section (head, body, limbs) can be rotated individually, according to preplanned timing. It can also be painted, and provided sections of the character do not change too often, they do not need to be painted or dressed for each frame, scene or section. There are several synthetic materials today which will stay in position for each single exposure. Consequently, a puppet figure could be built with wood for the body, malleable plastics for the limbs, linked with a ball-joint mechanism, and dressed in silk, lace or cotton with changeable features for eyes, nose and mouth. These were the main characteristics of the principals in Jiri Trnka's classic puppet film *Midsummer Night's Dream*.

It may be useful to examine here Jiri Trnka's achievement with his puppets (1912–1969). He was able to show certain dramatic and emotional effects with his figures that had never been achieved previously, and could only be seen in large-scale classical drama (*Prince Bayaya*, 1950; *The Devil's Mill*, 1951; *Old Czech Legends*, 1953; *Passion*, 1962; *The Hand*, 1964). He has also created some of the most satirical films (*The Good Soldier Schweik*, 1954), proving that it is possible to obtain a wide range of expression with puppets. Unfortunately, with *Midsummer Night's Dream* he may have chosen a subject which was beyond the capabilities of the puppet medium, being unable to convey the poetry. Nevertheless, he has proved that puppetry can handle the comedy and the spectacle of the original play.

Lighting problems in a puppet film are more similar to those of live action than to flat animation, where illumination is confined to a fixed surface. In most cases a three-dimensional set requires overall lighting with an extra spotlight highlighting the moving characters. If there is a great deal of motion on the set the spotlight should follow the path of movement across the scene. Whether internal or external, the style of lighting can contribute much to the mood of the scene and it should be as much a part of the preplanning process as camera movements. In dimensional photography both lighting and the camera can be controlled with much freedom (more than in flat photography) by viewing the set through the camera lens and changing the lenses according to the nature of the shot. Depth can be emphasized, with foreground objects nearer the lens, to enhance the perspective of a scene. The position of the camera is especially important. As a rule, the lens views the set from an angle of 45°. With computerized control it can change its position diagonally, and can track and pan, providing a degree of mobility beyond the capability of a fixed rostrum camera. It is advisable to work with a camera operator, if you can afford it, in order to concentrate on the animation of the moving objects and characters, which demands the maximum concentration on the part of the animator.

Trnka's success was not only that his work was based on the tradition of Bohemia and rooted in the tradition of Czech puppetry. Other nations, such as Hungary, Russia, Lithuania, China and Japan, have similar traditions. Trnka was also talented in areas where puppetry can be enriched, such as design, painting, innovation of character, sense of shapes and stories, and overall stage management. His disciples in Czechoslovakia, Pojar, Pixa and Barta, inherited his talent but not the degree of showmanship that Trnka brought to his feature films. One may consider his work as a climax of this genre, which will be extremely difficult to surpass.

Considerable achievements have been made in other fields of puppetry by two fellow Czechs, Karel Zeman in space fiction and special large-scale effects, and Hermina Tirolovain, with a contrasting style of handling small-scale, delicate but extremely difficult materials, such as wool and textiles.

The remarkable precision of animating hard wood belongs to the Hungarian George Pal, who took infinite pains to shape his characters' body and features, heads, eyes, mouth and ears for each frame of film according to preplanned calculations. His prewar films *Space Ships* and *Jasper* are classic examples of this technique. When he became the producer of Hollywood's first successful science fiction films *Destination Moon* (1951) and *War of the Worlds* (1953) he utilized his experience gained from his short puppet films.

Kihachiro Kawamoto, who produced *House of Flame* (1979) and *Dojoji Temple* (1976) in Japan and *To Shoot without Shooting* (1988) in China, is a true follower of the Trnka tradition with the addition of his own talent and the rich oriental style of puppetry of both China and Japan.

It is Co Hoedman who, in the West, has abandoned the traditional approach to puppetry and brought a fresh attitude to three-dimensional object animation. By using unfamiliar materials such as sand (*Sandcastle*, 1977) and transparent plastics (*Treasure of the Grotoceans*, 1980) he introduced a freer outlook to filming in three-dimensional movement as well as to the editing of his scenes. He soon discovered that motion such as walking does not look good on the screen so he tends to work in close-up. He knows how to make the maximum effect in plasticine animation and tends to compose his shots in depth and in diagonal arcs. He has a good sense of timing with three-dimensional characters and can manipulate them to act to the camera and achieve the correct movements with them. He also has the good sense to choose the type of subject most suitable for puppet animation.

Apart from Central Europe, the USSR and the Far East, the tradition of stop-motion puppetry is most evident in the studios of DEFA in Dresden and Berlin, where experts such as Katia Georgi, Gunther Rätz and Kurt Weil, among others, have achieved a high technical standard in producing series for children's entertainment for both cinema and TV. This studio has achieved what a puppet unit most requires: continuity of production with specialists in each separate department; dress design, set design, workshops, engineers, carpenters, three-dimensional animators, scriptwriters and composers all under one roof, and a public that genuinely likes puppet films.

Clay animation

More adventurous experimentation would further improve the content of so many talented puppet directors. Today, animating with clay is proving the most rewarding method. The main reasons are that there is practically no resistance of the material; it is easy to mould, bend and stretch; and it will stay in its place (provided one uses the right type of mould and keeps it at the correct temperature, since clay tends to melt under heat and become brittle under cold conditions). Many times, especially overnight, clay characters have disintegrated beyond recognition, obliging the maker to start again. Once the consistency test is passed (and it is not difficult to persuade the manufacturer to supply the right plastic clay with the correct resistance and colour), clay can be the most productive material to handle compared with wood, glass and perspex. This modest tactile substance is obedient to touch provided it is treated with respect. Following its introduction over twenty years ago its use has spread in many countries. At present, thre is a tendency to use it to create metamorphic transformations, due to easy assembly and disassembly of the material. There have been some remarkable achievements with it in the USA, the USSR and Hungary. Will Vinton in Ohio, USA, created *The Great Cognito* (1983), based on an impersonator's act. Vinton proved that with clay animation one can overcome the limits of live performance. With his wit and animation skill he made a hilarious satire on the world's leading politicians and showbusiness personalities, emphasizing their demagogy and exhibitionism with rapid transformations of their features. The effect on audiences was overwhelming. Once one recognized the character, the distortions which followed became pure fantasy. It re-enacted the early magic of cinema. Vinton, having full confidence in the clay material, named his organization Claymation, and ventured into the production of feature-length animated puppet films. He made *The Adventures of Mark Twain* (1985), a wholly praiseworthy effort to utilize the technique of three-dimensional animation. Without doubt, he discovered that clay and plastics were just as successful in a long animated film, in spite of additional technical difficulties in technical execution and artistic discipline.

One of the animators who understands the behaviour of clay is Garri Bordin, from Soyuzmultfilm, Moscow. In his film *Break* (1986), he tells a simple story which shows the technique to great advantage. It is about two boxers who make good use of being able to stretch their arms. They stop the fight and join forces to attack the referee instead. Bordin uses form transformations with humour and handles the material intelligently.

It appears that, so far, the most successful use of clay comes from Csaba Varga from Hungary, who has utilized the material to its full potential. He has created a character named Augusta out of a piece of clay, just a simple flat shape. As soon as she comes alive her features beome animated with gestures of Chaplinesque subtlety and expressions reminiscent of Marcel Marceau's mime. Her voice, a high-pitched sound effect, is in contrast with her delicate female character. Augusta has quickly become an international star. The situations she is placed in as an old benevolent aunt entirely suit her, such as making herself up to be beautiful, or preparing supper. She is constantly confused and everything goes wrong, whatever she does. Csaba Varga's humour, his sense of timing, his detailed observation of character, his story development and building of tensions are well within the nature of clay. The series shows what can be achieved with it as popular entertainment.

Paper in three-dimensional animation

It may have been due to the impact of *The Man Who Planted Trees*, a Grand Prix winner in the 1987 Annecy Festival, that the other Grand Prix winner sharing the honour was practically unnoticed. In retrospect, it was just as distinguished. It created much the same impressionistic effect with a totally different material, paper, a classic case of making something out of nothing. Here the paper was stiffened and made into an object to be animated. It was turned into a character, it had action and reaction, it had a range of motion, it was screwed up, rolled up, distorted beyond recognition, shaken, thrown away, kicked about but always recovered, resuming its shape. The maker, Boyko Kanev from Sofia Studio, Bulgaria, created a three-dimensional atmosphere to convey his statement about our disintegrating world. Appropriately, the title of the film is *A Crushed World*. This film, considering the slightness of its material, is one of the most innovative achievements in paper sculpture.

Paper is not quite as obedient as clay. It has to be tamed before it stays in a desired position. Figures can be constructed of wet paper fixed over a wire structure, which can be easier to control under the camera. The sets can also be made of cardboard and pressed paper objects, as in Kanev's film, to provide a stylistic unity with the moving characters. The lighting of the set and the photography can follow the same approach as any other three-dimensional puppet animation. It is interesting to note that Kanev and his team have had long experience in stop-motion photography, a technique applied throughout this film.

3 Computer animation

Computer films were first made in 1951, at the Massachusetts Institute of Technology, on a computer called Whirlwind. Such early attempts, however, had little impact beyond scientific and technical interest, and were confined to universities. The first major step towards wider usage was made by scientists at the Bell Telephone Laboratories in the USA in the mid-1960s; they rewrote the Fortran language for movie making, thus opening up the technique for animation. The last twenty years have seen a great advance towards the animator delegating the hard labour of animation to a machine, and we may have reached the half-way stage in the use of computers. Today, the technique, provided it is preplanned in a suitable program language, can reduce human labour, a condition most animators have dreamed of since the invention of cinematography. Furthermore, animated drawings can be conceived in three dimensions with textural complexities and richness which were not possible by hand. A detailed technical description of how it is achieved is outlined by such experts as Judson Rosebush in New York and John Vince in Britain, among others. A step-by-step description of the technology cannot be given here. However, the following is an outline of the basic process for a contemporary animator, with the understanding that if he or she wishes to specialize, a working knowledge of a system suitable for the task should be acquired.

Computer graphic systems are very diverse in configuration and capability. The range varies from a simple microcomputer of low resolution and low speed to such professional computers as HARRY. One thing is certain. No matter what technique one chooses, basic animation skills must be acquired and understood.

A digital computer is an electronic device whose functions are based on two values, 0 and 1. Instructions and data in this binary mathematical form (the software) are fed into the computer (the hardware) and manipulated as high-speed electronic impulses (as many as billions per second). To simplify the highly complex process, there are three stages:

Input: putting instructions into the system.

Processing: manipulating and storing the images produced, in the form of binary data.
Output: translating the images into visible form.

Input There are several ways of feeding information into a computer. The first two, the keyboard and the joystick, are generally limited to the input of very simple images. They can be used with editing software to produce more complex results, but this will usually require computer expertise.

The graphics tablet (or digitizer pad) is the most common and versatile input method, being able to cope with many different applications such as technical drawing or digital painting. It is now available in various sizes from 28 centimetres (11 inches) square to large sizes such as A8. Tablets are very easy for an artist or graphic designer to learn to use, and sometimes a learning period of only a couple of hours is required. The information generated is stored in one of two forms: either as individual points (pixels) or as point-to-point lines (vectors). Vectors are quicker and easier for the computer to manipulate, but some complex images are stored in pixel form because the vector method would prove too crude. Most software using this kind of input will use areas of the tablet as a menu, allowing the operator to select available options in the program. These areas can be indicated with a suitably printed overlay.

The digitizer pad consists of a white plastic surface which is fitted with a matrix of horizontal and vertical wires reacting to a low-level electronic signal generated by a touch of the stylus. The stylus also contains a small photosensitive cell which transmits a command, via the screen, at great speed. The digitizer cannot be connected to a computer without a software program to guide its output. The digitizer is a highly sensitive instrument, performing a highly complex operation and, for physical handling, is not always comfortable. The joystick is easier to manipulate. It is connected with springs and, when pressed in any direction, generates an electronic code to activate a computer program. The system functions from a comparatively simple menu of commands attached to the terminal.

The last form of input is the video camera. This is used to capture artwork or photographs which can later be modified by, for instance, a digital painting system. This is often called 'frame grabbing', and requires a video camera to be connected to the system. There are two main types of frame grabber available: those that capture a monochrome or grey-scale image, and those that can record full-colour images. Pictures that are input in this way are stored as pixels and can be altered using a digitizing pad and 'painting' software at a later date.

Processing In many computer graphic systems the computer itself is small and hidden among the various peripherals. On the other hand, some systems use very large 'supercomputers' such as the Cray. There is no general rule. The requirements of graphics applications are very different from those of a typical business system, although many of the same processes are used. It is obvious that a computer with a lot of processing power is required to handle the large number of repetitive calculations that most graphic software routines use.

The computer will have one or more terminals to communicate with it, probably disk drives and/or tape drives for storage as well as the more specific graphics peripherals.

The system's ability to manipulate the images it is storing will depend largely on the software being used. It is in this area where there is the most diversity and duplication, and many people may find themselves laboriously writing software similar to what someone else has already written. It is possible to generalize to some degree about currently available software. Some will operate only in two dimensions (2D). This is usually the kind that was developed for technical drawings and has been modified to cope with animation. Other types, also 2D, will work on complex 'painted' images (e.g. the Quantel Paintbox). These usually have some scaling and panning among their capabilities.

Other software can manipulate scenes in three dimensions, giving true perspectives and allowing the placement of imaginary light sources with appropriate light and shade. The more advanced of these can be made to produce images that are both very complex and very lifelike, if that is the final objective.

In all cases, animation is produced either automatically by instructing the computer to move an object or a viewpoint from A to B in a given number of frames, or simply by using the computer to help the animator to draw each frame and then to reproduce the frames in order. In some cases this animation can be viewed in 'real time'. However, this is uusual, as the images are too complex and have to be retrieved frame by frame onto the chosen output medium.

Output It is in the system's output that most changes have occurred over the last few years. The first commonly used output device was a pen plotter using paper or film. This method is still essential where working drawings are required or where cels are to be coloured in at a later date. Normally, the computer will have some kind of display processor (frame buffer) attached to it, which allows the images to be displayed on a video monitor. Partly because of falling prices combined with a great increase in performance, video display devices have proliferated.. There are two basic types of display available, vector types and raster types. As raster devices can, to a large extent, emulate vector terminals, a flexible system will probably use rasters. Today the raster types dominate the market.

The minimum display resolution that seems to be acceptable to the professional user is (in pixels) 512 by 512 or more. A typical middle-range display will use 768 by 575. This resolution produces very acceptable results for most applications and has the added advantage of being compatible with a PAL TV output. It is this, more than anything else, that has made the relatively low-cost system a reality.

Having generated the image you want on a colour monitor you now need a method of permanently recording it, and here there are three main output options: plotted, film or video. The output of a pen plotter using paper or film suffers from being rather slow and inflexible in operation. It has, however, been used extensively to produce animation sequences. Often the originals are coloured in by hand.

Film, both 16 mm and 35 mm, is the most universally acceptable output medium owing to its high resolution and wide use in all parts of the visual media. The cheapest way to get a photograph from a computer display is to place a camera in front of it and take an exposure. However, while this will produce a result that may appear passable, it will suffer on account of scan lines on the surface of the digital monitor, commonly known as RGB (red, green, blue). The solution is to use a purpose-built camera system which is designed to smooth out many of the inherent defects of any video monitor. There are two major types available: relatively low-cost analog systems and high-cost high-resolution digital ones.

Analog cameras use the RGB monitor outputs from the graphics processor and display the image on a high-resolution monochrome monitor in three passes, one for each of red, green and blue. This method overcomes many of the problems previously mentioned, as monochrome monitors do not use a shadow mask, and, by modulating the horizontal scan lines, the image can be made to appear perfectly smooth. One advantage of these systems is their speed. They can usually shoot a frame in 20 seconds, but they are limited to the resolution of the display system being used.

Digital frame records work in a similar way except that the image is transferred digitally from the computer to the camera and the resolution can be 4000 lines or more. Typical shooting times are in minutes per frame.

The film camera used will depend on the film format required, the most common being 16 mm or 35 mm for animation or 35 mm for stills. The standard animation camera can often be modified so that it can be operated automatically by a signal from the computer when the data for a frame has been processed. This method means that the shooting can be done overnight, freeing the system for other work during the day.

Video is an increasingly popular output method because of its speed, flexibility and cheapness of operation (although some of the hardware is expensive). Video output is derived from a computer by coding the RGB outputs of the display processor in one of the recognized video standards (PAL, NTSC or SECAM). This can then be recorded, edited or post-produced using any of the techniques currently available.

Once the computer has done its job, one can resume the normal animation steps with the synchronization of soundtrack, editing, dubbing and other stages needed to complete the film. It should be realized that computers can now be utilized in several ways:

- As a principal tool in the actual animation process;
- As a part of the process to assist in:
  (a) Ink and paint systems;
  (b) Video recording systems;
  (c) Motion control, graphics and models;
  (d) Interactive video disks;
  (e) Video special effects systems;
  (f) Rostrum camera motion control;
  (g) Video line testing;
  (h) Laser projection.

With animators in mind, the computer hardware industries produced the turnkey system a few years ago, i.e. a computer hardware and software system combined as a single unit. The name turnkey is derived from the fact that with a key, all one needs to do is start up the system. There are several kinds of two-dimensional animation (such as colour cycling, paint, typographic and drawing capabilities) on the one hand, and three-dimensional modelling, shading, texture imaging and wire frame previewing on the other. Here, the system can have a large paint range and build three-dimensional objects in space by combining polygons. Both functions, as a rule, have complete two- and three-dimensional capabilities.

One can expect that real-time operation in colour will be

incorporated into all future turnkey systems and eventually users (among them, animators) will require fewer operating skills and less training to use them, but it will be some time before the two functions can be combined. Logic may not be the best guide to such a marriage, despite the fact that the animation industry desperately requires modernization. It could benefit from saving time and money by eliminating the drudgery of inbetweening, tracing and opaquing, which have handicapped it for over 70 years. It is evident that the most serious obstacle to faster integration between the tools of the traditional animator/ craftsman and the electronic system is fear, a resistance to anything connected with machines and technology. This 'technophobia' can only be cured by time and persuasion to prove that it can be in the animators' interests to broaden their skills. However, the most important factors which animators have to comprehend is that they need not lose artistic integrity, and that the computer is an instrument which does not take control. When such a situation arises the creator will be able to handle the computer with the same ease as a pencil and brush.

Microcomputers

One of the most significant developments has been in the area of home computers, since it involves millions of people experimenting with their skills in their own homes with their own equipment in their own time. Fully programmable microprocessor-based interactive games have become a very large industry with practically all the major electronics companies competing with each other. This, unlike the optical film industry, has resulted in dozens of systems, but few of them are interchangeable.

The complexity and the scale of microcomputers based on the silicon chip with its complex integrated circuitry has become the subject of hudreds of specialized magazines, with a circulation of several millions, announcing new machines, new electronic software, new forms of electronic games (e.g. *Dungeons and Dragons*) packed into cartridges and disks.

The latest newcomer for the Japanese Mitsubishi MSX computer is the Sanyo speech cartridge. All the speech-generator circuits are built into the cartridge. Sound is output through a TV speaker or a hi-fi system connected to the computer audio output. The text is in fluent English. It is used for educational games, such as teaching reading, writing and the alphabet and also announces words such as green, yellow and red when the appropriate colour is shown on the screen.

The first game screens were designed mostly by programmers. Today, graphic designers and animators are increasingly involved in the process. Obviously, both the professional designer and the animator must be instructed for their work to be suitable for putting onto a computer. Most video game designers have the skill of being able to sketch with a stylus directly onto an electronic screen. Since the market for well-conceived games is very large, there is a choice of video games to be used with a large number of different home computers. Many Hollywood studios have specialised in this particular field, which offers great scope for the work of professional animators.

It is primarily in the three-dimensional aspect of computer graphics where new types of visual concepts can be created and fantasy and creativity are most likely to succeed. The concept of *ray tracing* is an example. This three-dimensional modelling technique has created some remarkable results which only computer graphics can produce. Ray tracing can be compared to sculpting an imaginary world within a confined space, providing shapes, textures and surfaces not seen before. However, it is a very expensive process, which takes up much computer time. It requires the involvement of supercomputers, which are not easily available.

Naturally, most computer graphics work is carried out with microcomputers and a simple keyboard. It is, however, a common error by thousands of micro practitioners to assume that micros can perform like a Cray supercomputer. The 'horses for courses' concept has never been more true than in this situation. The input will always determine the output, and one should never expect more than a system can provide. At the other end of the scale, one has to justify the use of a constantly expanding technology and fully realize its potential as a contemporary tool to express new ideas.

In spite of all the difficulties, interaction between the developing market and new electronic picture-processing tools is inevitable. They are interdependent, artist and technology together. The animator can only benefit by learning new methods of communication. With animation's creative talent adapted to new tools and services, many fields can be opened up in education, in the wide range of commerce, in new applications of design and in entertainment. If such an objective is successfully achieved, animation could broaden its services to all levels of the community.

4 Market potential

Screen Digest, a monthly magazine specializing in the analysis of the visual communications industry, reported in its June 1989 issue that animated cartoons were 11 per cent of the product used for children's entertainment in television worldwide. This is much smaller than one may have assumed, but statistics have shown that children's entertainment is a small proportion of the total television diet in any case, and that most of it is produced in live-action format. Since TV series for children are a substantially larger proportion of the world's animation output, one may come to the conclusion that the TV market is very large indeed, and that the volume of animated series for children could be increased at least to double its present level. It seems that this type of activity does not attract newcomers at all. There certainly is no rush of animation studios wanting to specialize in this genre. Newcomers tend towards units producing experimental films and advertising spots and studios specializing in computer graphics. But the field is wide, and is growing all the time. The main markets presently are:

- Commerce
- The leisure and entertainment industry
- Titles, inserts and excerpts for television
- Full-length animated feature films
- Entertainment series for television
- Educational films
- Science
- Public relations
- Architecture and industrial graphic design
- Experiments

There is a wide option both for those who seek employment and for those who work on their own or in small groups. The following are markets to explore.

Commerce The main activity under this heading is advertising production for television and cinema. For the animation industry this has always been a prime source of revenue. From the beginning, the freedom of animation, its capacity for rapid change, transition, simplification, humour, penetration beyond the surface proved to be an asset for the advertising agencies responsible for commissioning such work as well as for clients presenting their products to the public. In the pre-television era, advertising cartoons used to be five minutes long, containing a storyline strongly related to the product to be sold. As time passed the films became increasingly shorter, down to two minutes and then one minute in length. Animators were required to adapt, and tell the five-minute story in one. They had to compress, edit and use shorthand visual symbols to make their film work successfully. Today the length of a commercial may be 15 seconds. For such a discipline a new visual language had to be adopted to compress the meaning of a spot and make it comprehensible to mass audiences. Few artists have the skill required to achieve this. As a rule, they are well rewarded, and gradually TV spots have become the most lucrative market for the animation industry.

The leisure and entertainment industry

Under this heading there is the production of animated titles for expensive feature films (the James Bond titles, by Maurice Binder, are an example of the type), as well as the special effects in science fiction films, with budgets being most generous when compared with other types of stop-motion activity. These animated inserts have contributed a great deal to the character of the genre and have opened up a new market for animation using the newly developed computer technology and stop-motion remote control. It has helped to develop digital computer animation which, since its early uses in the *Star Wars* films, has been utilized in a number of other fields such as scientific research and space technology.

Titles, inserts and excerpts for television

This market has only recently developed. It is at the point of rapid expansion due to the lively contribution an animated opening can provide to programmes. Until recently, most TV companies depended on their own internal graphics department. Some, like BBC TV, contained a staff of 120 personnel (animators, designers and electronics technicians), but it has been proved that the freelance supply is no more expensive and is more imaginative, with a much greater degree of style and graphic invention. In work commissioned by TV, much has become a catalyst for combining new technologies with more traditional ones. The short length of the films leads to many experiments. The problem in this field is the small budget, the time pressure for delivery and the fact that the station will repeat the film hundreds or even thousands of

of years can be compressed into a few minutes of screentime. A mathematically complex theory can be visualized graphically, illuminating the main points of the argument. An animated film's ability to simplify visual information can provide better memory retention than a written text or oral lecture. There are further advantages in classroom situations, where live-action films used to dominate the situation:

- The film can make use of live-action, and then break away from it to animation when points need analysis or clarification.
- A moving diagram can be superimposed on an image of actuality, and in so doing it can simplify (and analyse at the same time) the working principles that may not be at all clear in the live-action film.
- The basis of animation technique is the superimposition of transparent cel sheets one over the other, each bearing one section only of the final composite moving image. In this way, the sectionalizing of a complex movement (for example, a part of the human body) can be stripped down by peeling off layer after layer of the total image, so that the whole may be clearly seen as the sum of the parts.
- Animation can introduce every variety of graphic style and presentation that the animator can devise to illuminate the principles the teacher has to demonstrate. These can range from wholly imaginative or symbolic images to the simplest diagrammatic representation, according to what is most appropriate. Therefore aesthetic qualities can be added to purely intellectual ones.
- Included in the graphic style and presentation are the advantages of colour. Colours can be used to help convey differences and distinctions. They convey by eye alone what may otherwise have to be laboriously spelled out in words. Colour, too, can be used symbolically in the kind of subject which has psychological overtones or involves mood or feeling.
- Animation can, more flexibly, conveniently and with greater variety than in live action, reduce or increase speed in the presentation of movement in any process which is being demonstrated. Important elements may be presented in slow motion, while the unimportant can be speeded up or omitted.
- The soundtrack can be exactly timed in animation so that comment, whether by words, natural sounds or musical counterpoint, can be synchronized, if need be, to one 1/24th of a second. The use of natural or artificial sound can help to reinforce the image.

In spite of these advantages, animation has not been sufficiently utilized in teaching films. It is still waiting to be explored. Since many classrooms are equipped with video playback, it would be easier both for the teacher and the pupil to use animation on video. Here is a market to open up for the benefit of society, especially in underdeveloped countries.

Presently there seems to be a serious deficiency at the junior level of scientific education. Provided productions are related to the science curricula in schools, there is an untapped market. Sponsorship on a large and imaginative scale is needed here both by the educational authorities and industry.

Science

Most animation production in science is in the computer departments of universities by scientists for their own use and as an aid to lectures. Some of this material can be quite brilliant. The *Galaxy* tape at Syracuse University, New York State, is a case in point. This is a research film showing the movement of stars during a period of 200 million years, with the formation of a galaxy as the climax. The length is 3 minutes. The work is totally satisfying, both in motion and as an aesthetic experience. However, it is most unlikely that the tape will be seen outside the confines of the university.

There are many areas where science and art meet in research and development departments of subsidized institutes which cannot receive close attention in the commercial market. Soft-edged changeable matter such as smoke and water, reflections, shadows, sine waves and textures are examined and programmed to find out their behaviour. Compared with European institutes, those in the USA such as Ohio, Stanford and Berkeley are rich. So are Osaka and Hiroshima universities, where such valuable groundwork is taking place. Most of them have powerful computers like the Cray X-MP to assist their research, which could eventually reach the commercial market but in a much simpler form. It is unlikely that without basic experiments in universities such special effects as those in *The Last Starfighter* and *Return of the Jedi* would ever have been realized.

Public relations

Animation can make entertainment out of facts, figures, systems and ideas. If presented in forms such as printed brochures, circulars or live-action documentaries, these subjects can be dry, uninteresting and uninspired. Animation can be quick, amusing and, if it is well conceived, crystal clear. Classic examples have been films made by the Larkins Studio, turning the dry figures of the balance sheet for ICI shareholders into a highly amusing, witty film explaining the company's finances. This was an excellent public relations exercise. The public's goodwill was quickly won, and it was ready to invest more money in the company.

Public relations films are possibly more widespread than is realized. They are not only used to obtain public goodwill for a company but also internally to inform, educate and generally communicate with staff. A series of films was produced, for instance, by the American company, Dupont Incorporated, to explain the workings of its staff pension scheme.

Public relations films are not confined to industry. They are used by governments and agencies such as the World Health Organization in Geneva, to combat drugs and alcohol abuse. The United Nations has also sponsored films with a universal message. Bratislav Pojar's *Boom* (1979) was a successful anti-war film. The author's own film *Doctor in the Sky* (1984) was sponsored to persuade member nations to stop arming themselves. Jonathan Hodgson and Susan Young's film *The Doomsday Clock* (1987) for the United Nations has the same objective. One of the latest examples of public relations films is Bob Godfrey's *La Belle France* (1989), a humorous adaptation of the story of the French Revolution in 1789, sponsored by the French semi-official organization INA (Institut National de l'Audiovisuel). These films are delightful entertainment, making a point painlessly. Most of them are winning the public's goodwill and sympathy for issues which otherwise could be disregarded, ignored or opposed. The impact of these productions with their charm and force of presentation makes them popular with sponsors, and it is most likely that they will continue to maintain their importance in the market.

Architecture and industrial graphic design

As with the production of science films, in these fields animation is basically an in-house activity with microcomputers. Since there is a desperate need for competent assistants, it is an area where it is comparatively easy to find employment. The task can be interesting. Like a special effects animator, an architect spends a considerable time in building the basic elements of a structure. This could be a whole new city or road system or just a single house or bridge. It is far more economical to animate computer models of these objects in order to examine them from all angles and make adjustments before the physical construction starts. The amount of saving in labour and installation could be very large, since it is possible to modify the structure as one develops the design and models.

Many organizations have their own graphic workstations, keyboards, monitors and digitizing tablets for input devices. In this way, a unit has access from energy analysis to instant two-dimensional processing. The mechanization of the whole system can include such details as electrical installation and window placing. There are a number of specially prepared software systems for architects, and one can see that it is an activity which has changed considerably in the last few years.

So has industrial and graphic design. There is no longer any need to build a car or aircraft before it is tested. Computer animation can test from all angles, including the positions of drivers, pilots and passengers. New industrial products can similarly be examined. These tryouts save manufacturing time on a project and today have become an accepted routine in industrial companies.

The latest technology is desktop printing, an activity carried out in graphic design studios. This combines design with computer animation to produce printed material more precisely and quickly than with manual systems. In this field, as in others, the opportunities of employment are high for the right students who have, or are prepared to learn, the specific skills.

Experiments

This activity has been the most exciting field for animation since the art was invented. Experimental films cannot be categorized. They can be made by beginners, experienced directors, students or just enthusiastic amateurs. Most of these people are motivated by a desire to widen the artistic and technical aspect of the medium. When the history of animation is written it will concentrate on such individuals who had the capability, tenacity and talent to fight for their ideas against ignorance and indifference. Every decade since the time of Méliès, Winsor McCay and Emile Cohl can tell a tale of individuals whose aim was to advance animation from their temporary bases. In the past 20 years the rate of experiments has accelerated due to the revolutionary technological changes in electronic animation. The main areas to mention are:

- Mechanical development
- Exploring the impossible
- Art in movement
- Broadening the animation language
- Computer graphics

Most of the pioneers' work was carried out without an existing market, but many opportunities have been opened up as a result of their experiments, especially at the beginning of cinematography, when everying that moved was magic.

Mechanical development

Once the creation of the illusion of movement had been established by Eadweard Muybridge and E.J. Marey as a scientific reality, and Edison and the Lumière brothers had made a system of projection possible, pioneers such as Emil Reynaud, Méliès and, later, Winsor McCay and Emile Cohl explored the medium in public. The public was fascinated by such a novel device as stop-motion photography, and flocked into cinemas to see it. The market was established and ready for expansion.

Exploring the impossible

The most effective way to please the film maker and the public was to speed up the action, and make the characters somersault, compress, elongate and transform their shapes as well as defy the force of gravity, which was possible only with animation. The early period of live-action films followed the same formula of speeded-up action, and burlesque and comic live-action shorts appeared.

Art in movement

Major movements in art such as Impressionism, Expressionism, Futurism and Cubism related to the birth of the cinema at the end of the nineteenth century and at the beginning of the twentieth. Many artists experimented in both media, Léger, Duchamp, Alexeieff, Richter, Fischinger, Len Lye and McLaren among them. Basically, the two media remain different. Static art depends on its immobility, confined to fixed dimensions, while the moving picture opens up extra dimensions by the inclusion of motion in space, passage of time and sound. For a time, when no-one could successfully resolve the basic differences, film-makers tried to achieve mobility through the movement of the camera. Some still believe that there are limitless possibilities in combining static paintings with fluid animation. There is almost 80 years of experiment that has never been explored properly. The work of pioneers such as Luciano Emmer, Manfredo Manfredi, Pavao Stalter and Peter Foldes is waiting to be recognized, with animated painting specially created for the camera and with paintings texturized and originally designed in continuous animation.

Broadening the animation langage

When comedy was the major objective of animation, the majority of cartoonists had a simple formula of expressing themselves through gags. It worked. For 50 years the genre has employed thousands of experts. Their skill and inventive artistry today is appreciated as a golden period of animated cartoons. The films of Tex Avery, Bob Clampett, Max Fleischer and Pat Sullivan became classics. The formula was based on conflict between two characters, one small, one large, with continuous physical confrontation between them, quick recovery and repeat action.

During this process it was essential for the small character to outsmart the large one, and for the large character to suffer the consequences of his bullying. While gag cartoons exist in great depth as the genre (to the extent that the majority of audiences mistakenly know of no other type), it leaves all other animation genres with the difficulty of establishing themselves. Animation is still waiting for greater recognition in satire, parody, poetry and lyricism, drama and surrealism. Most of these have an adult appeal with high visual content. They contain artistic and graphic design qualities of an experimental nature and a type of new experience for the public. In spite of hundreds of animation festivals and high-quality achievements by many contemporary artists, there is an element of resistance in the market to recognize these types of genuine entertainment.

The pioneers of adult animation made their contribution soon after World War II through UPA (United Productions of America), a splinter group from Disney, where John Hubley, Bob Cannon and Steve Bosustow streamlined their characters into modern structural forms and started to bring the work of modern authors such as Ludwig Bemmelman (*Madeleine*) and James Thurber (*Unicorn in the Garden*) onto the screen. These stories contained satire, visual wit and a fresh graphic style with a strong adult appeal. In spite of their success, after a decade the unit disbanded but left an influence which still dominates the USA to this day. In Europe, where the gag tradition never became so deeply rooted, studios in Yugoslavia (Zagreb Film) and Britain (Halas & Batchelor and TVC) adapted their styles more easily and made their contribution to adult visual language more rapidly. With the full-length animated film *Animal Farm* (1954), based on George Orwell's classic novel, a sense of poetry and subtle parody was introduced with an adult appeal not previously experienced with other long animated films. George Dunning's *Yellow Submarine* (1968), with music by the Beatles, was also an adult experiment, opening up new prospects in graphic design.

Today the field is crowded with talents whose style and thinking would appeal to adult audiences. Two particularly stand out: Frédéric Back, the French artist working at the Canadian Broadcasting Corporation (*The Man Who Planted Trees*, 1987), with his fluid classical animation, injected high-quality individual adult language similar to the style of Renoir; and Yuri Norstein, with *The Tale of Tales* (1979) and *The Overcoat* (1990), proved that there is no limit to fluid tonal animation in three dimensions, even without the aid of a computer. Through these achievements and so many others it is not difficult to prove how rich the potential of animation can be and what an exciting prospect a newcomer could have. However, the new generation of animators will have a difficult task to break down resistance and to teach the public to enjoy and understand this new animation language.

Computer graphics Experiments using computers are taking place at both extremes of the computer graphics field. On the lower level, a simple Macintosh or IBM PC can be used at home for pleasure. At the other end, artists are using powerful computers as experimental tools, mainly for the computers' interactivity and ability to create three-dimensional perspective projections in space to change, store, erase and repeat images. Students who have been used to flat manual animation find it difficult to adjust to electronic systems. Brian Jenkins, a computer animator at Kroyer Films in Hollywood, who co-directed the computer-made *Technological Threat* (1988), has suffered through the transition. He maintains that it takes two years to understand how a digital computer works and another two years of practice before one is able to make a scene. Since he is an intelligent and talented person and well motivated to absorb information, one can assume that others might take much longer to become acquainted with the technology.

The field of computer graphics is enormous, with a large range of software available, but there is no compromise in this field as there is in hand-drawn animation, where one can change one's approach. You either understand how your system works or not. If not, then do not touch it. If you do, it is most likely that a new world has opened up for you and you will never wish to turn back again. The same principle prevails between live-action photography and hand animation. If an action can be done better with live action, do not employ animation. Animation starts where live action ends; it is the difference between the surface and beneath the surface, where realism is left behind. Computer animation repeats the same concept. Use it only when no other technique can be employed. The menu is plentiful, with automatic repetition, mechanical duplication, simulation of dimensions and perspective, three-dimensional modelling, texturization, complex special effects and speed of action at a breathtaking rate. There are more opportunities as new equipment appears that is easier to use, carrying out one's commands automatically.

Service studios

Large-scale service studios started in Japan in the 1960s, when labour costs were still low. This is no longer the case. In time, Japan, like the USA and Western Europe, has moved towards co-production with nations with lower living standards and wages. Once, such services depended on motor transport. In South

Korea, busloads of animation were sent around to young mothers to trace and were collected next morning. This service proved unreliable, and eventually stopped. American studios found that unless they applied strict supervision over the quality of tracing, unacceptable distortions occurred, and in painted cels colours did not match. When these difficulties were solved, local labour was used for animation based on layout and workbooks prepared by experienced animators from the country of origin of the commission. This method worked better. In spite of the many difficulties, activity is spread worldwide. Disney uses studios in Taiwan and South Korea, Hanna Barbera has expanded into Spain and Australia. The highly distinguished French director, René Laloux, has just completed a feature-length film of 83 minutes (*Gandahar*) in North Korea in the studios of Sek de Pyongyang. It appears that international communication transcends political differences when it comes to labour costs.

5 Animation mechanics

Both hand and computer animation require knowledge of how animation works. At first, the purpose of animation was to make an object or figure move in stop-motion with spaces equally divided from one key position to another. Experience taught that, by various means of slowing or speeding up the motion, one can inject character and a sense of rhythm into the movement. In time, the process acquired sophistication to such an extent that animals could behave like humans (Mickey Mouse, Felix the Cat) and humans could be caricatured (Popeye, Betty Boop).

Today, after the disappearance of the cinema market for short-length animation and the emergence of television, there is a totally different structure for animation. There is no longer a need for highly sophisticated and refined movement; television's small screen would not be able to convey the artistry. Markets and budgets are different; one speaks of limited, semi-limited and full animation. Lip sync movement can be carried out to limited and not so limited extents, and most animators are concerned with the creation of illusion, where the yardstick is what one can get away with. New skills emerging are those of constantly manipulating the number of drawings used within a second of airtime, and the length of holds a figure should stay motionless in a static situation. It is this aspect of animation which is being examined, perhaps for the first time in such detail.

Let us consider the extremes. First, if we expose a drawing repeatedly so that it is seen for a full second (24 frames for cinema and 25 frames for European television), we appreciate such a drawing in its own right, like a frame in a comic strip. Second, if we expose a series of drawings at a rate of one frame per drawing, they become an animated sequence, to be appreciated for its movement. They would not be seen any longer as single drawings but judged only on how they relate to the other drawings before and after. The separate segment has become a unit of time.

What is full animation?

Full animation can be divided into two separate but partly overlapping aspects. First, there is the problem of characters and objects moving in a convincing way, based on the laws of nature.

Characters and objects have weight, because they are acted upon by gravity. Second, there is the problem of making characters move in an expressive way – creating moods and feelings which we recognize as being related to the way people, animals, etc. behave in the real world. This is acting. It includes mime, gestures, facial expressions and reactions.

In full animation, economy may not be the first consideration. The priority is to give a feeling of credibility to the characters, to make them look alive, with thoughts, muscles and weight – moving about in space in a recognizable way. The following are a few points for consideration:

- Do not try to copy reality.
- Make a study of actual movement – but *simplify* it, extract the elements you want from it and exaggerate them to produce the maximum dramatic effect.
- Give the character time to *think* where necessary.
- Where a particularly smooth effect is required on a large screen or where animation has to be fitted to live action – or in some camera-table or peg movements – single-frame animation must be used.
- Single-frame is also used when double-frame animation is too slow to explain the action. For example, a fast run cycle of four frames to a step would obviously not work on twos.

What is limited animation?

This is a method of producing entertainment cartoons for television series, and also instructional films, economically and quickly. When we expose each drawing for two successive frames, considerable smoothness of movement can be produced. This is partly for economy and speed of production, and partly because many effects work acceptably on double-frame, without having to resort to single-frame. Films made for television, for instance, are almost entirely shot on twos or longer.

Preplanning and layout are of extreme importance. Much production time and work can be saved at this stage by planning the continuity to avoid complicated scenes, such as smoke, water and cloud effects, and perspective animation. Some ingenuity here can save work later.

Limited to full animation depends on how long one holds a drawing. The action becomes smoother as more drawings are used. The usual convention is to use double-frame animation, one drawing for two frames. If one drawing is used for three or four frames, the animation will appear progressively jerkier.

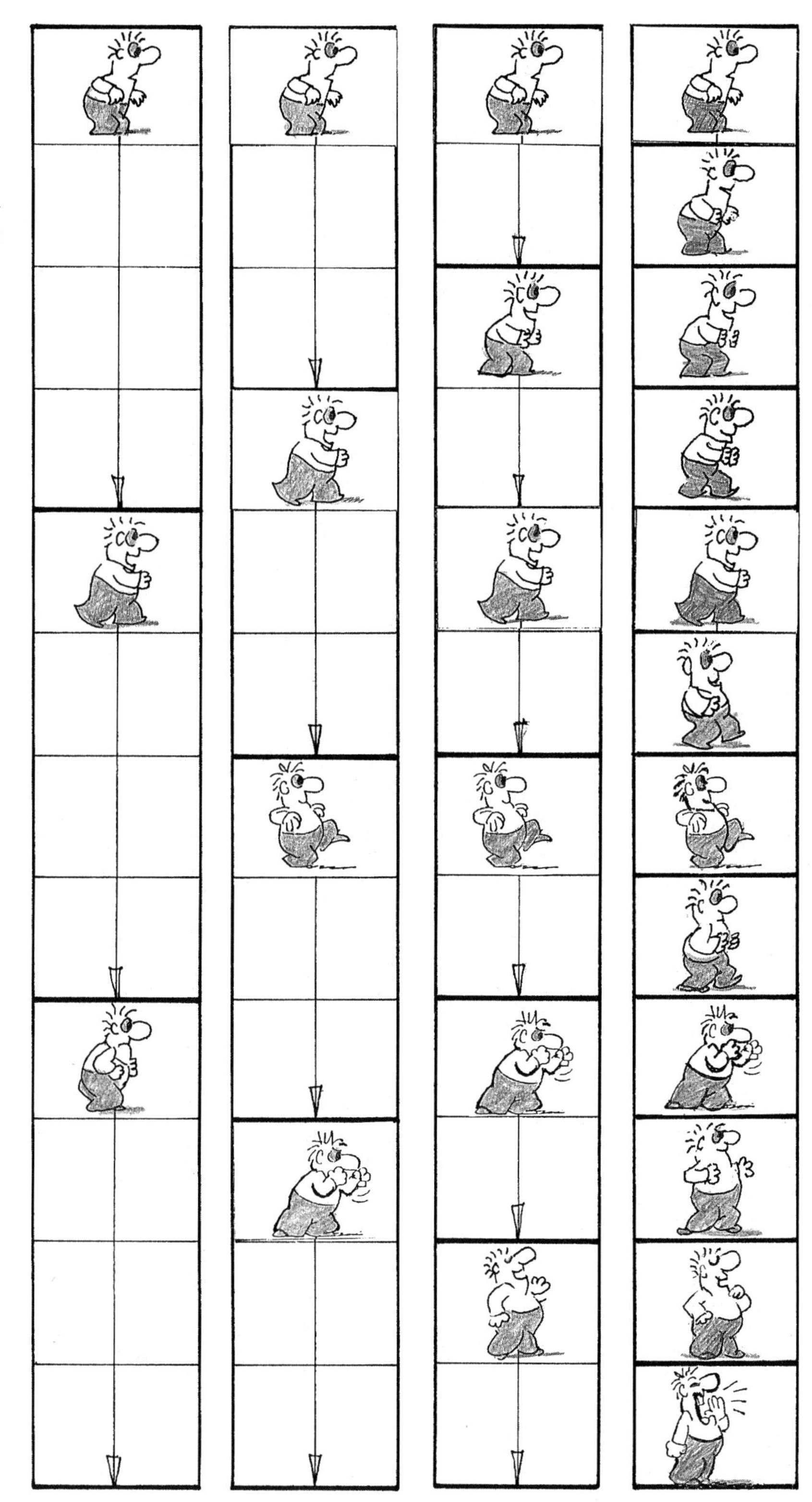

Animation keys drawn by Borivoj Dovnikovic, the Yugoslav animation director, who is a master of being able to reduce his characters to their simplest form in order to be able to maximize their expression and gestures.

A bitter argument amicably settled. This complex situation is conveyed by Borivoj Dovnikovic with basic simplicity of lines and animation effects.

In limited animation, animate as little as possible to keep the movement flowing. Separate as much of a character as you can onto hold cels. Use separate arms, heads, eyes and body, as far as you can. This will also help the trace and paint department.

Try to arrange the action to use short cycles where you can. For example, a character might be doing something in one place – 1,2,3,(1),2,3,(1),2,3 – then moves to another pose – 4,5,6,7,8, and does something else – 9,10,11,12,(9),10,11,12,(9), 10,11,12,(9), etc. Try to keep the action clear and interesting.

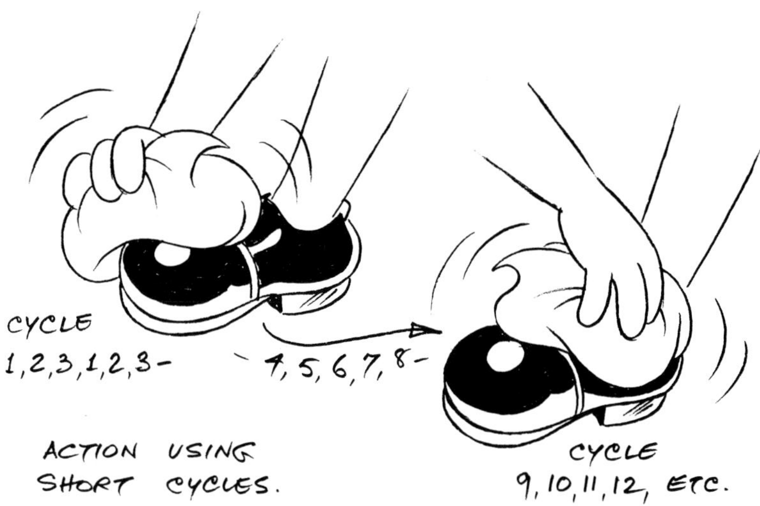

In spite of the fact that walking cycles do not contribute a great deal to the interest of a film, they are constantly used. Some entertainment value can be gained by making animation cycles as characteristic as possible. The walk can reflect the look and expression of a character, breaking away from the two-dimensional flatness of movement.

NORMAL

SWANK

DEJECTED

On a walk cycle it may be possible to separate the body from the legs so that one up and down movement of the body fits both left and right steps. This approach economizes on the number of cels used and on tracing and painting.

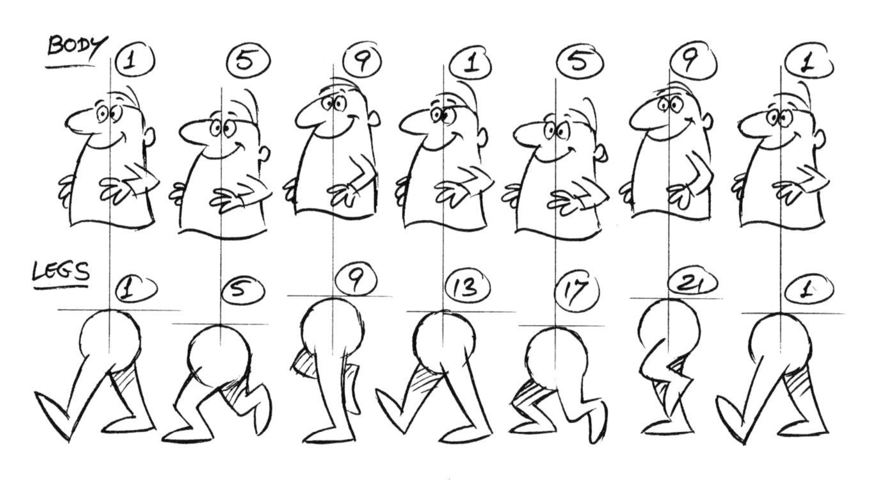

LIMITED WALK CYCLE

PAN BG.
MEDIUM CLOSE-UP TO AVOID LEG ACTION
CUT AWAY FROM COMPLICATED ACTION TO A CHARACTER'S REACTION TO IT.
B
A
(A) C.U. MAN REACTS TO DOOR OPENING EFFECT (OFF).
(B) TRACK BACK. - THE SECOND CHARACTER HAS ALREADY ENTERED.

A film with limited animation may need to be relatively strong on background and design to hold the visual interest. Keep characters simple and avoid things like floppy clothes, tails, etc., which call for extra drawing. By cutting to a medium shot of a character walking (waist up), leg animation can be avoided altogether.

There are occasions when an action takes place outside the film frame, without the audience being conscious of it. On these occasions it helps if the sound effect can reinforce such action.

PAN BG. ON TOP PEGS →

MAN WALKS ON SPOT ON STATIC BOTTOM PEGS.

— USING SAME CELS AND BG. —

STATIC BG. ON TOP PEGS.

MAN PANS ACROSS ON BOTTOM PEGS ←

IF THE MAN IS TRACED AND PAINTED ON TWO SEPARATE CELS, BOTH CAN BE TURNED OVER ——

—— AND THE SAME CELS CAN BE USED IN REVERSE.

When a basic walk cycle has been made on long cels, much can be done under the camera with peg movements (walking into and out of screen, etc.).

There are a number of methods of animating a walk, but all consist basically of the way the character transfers his weight from the back foot to the front one and gives a forward impetus at the same time. There is usually an up and down movement of the body, with the front knee bending to cushion the weight of the body as it descends.

Remember that if a character decides to start moving, then, according to how heavy he is, it will take time and effort to overcome inertia and get his weight moving. Similarly, if he has momentum and wants to stop or change direction, this also will take time and effort. For a quick exit from a scene, animate the whole body as the character starts his weight moving from rest – perhaps his feet slipping until he overcomes the inertia of his body, which tends to stay where it is as he tries to move it. Any extremities will tend to follow slightly later (coat-tails, ears, etc.).

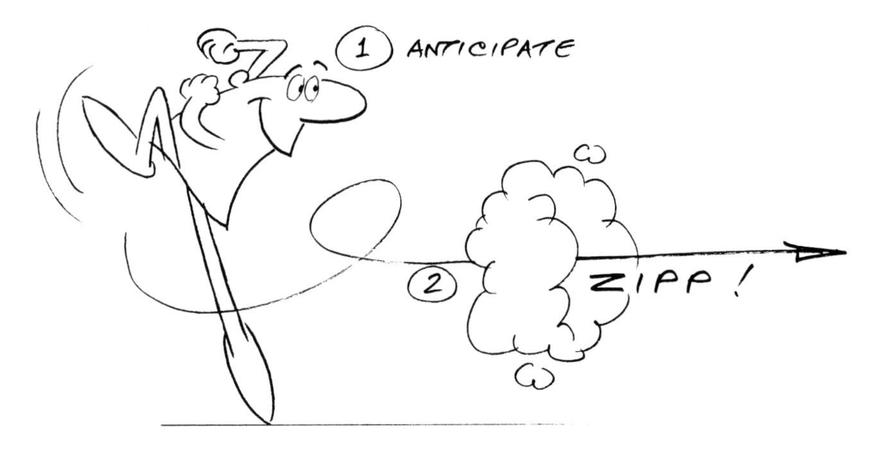

To get a character out of a scene quickly it may be sufficient for him to go into a rapid 'anticipation' pose (1) followed by drybrush effects and/or dust as he zips off-screen (2). The uses of after-action effects help the illusion of action continuity, but many animators tend to overdo it, which can be a handicap to a movement instead of a help.

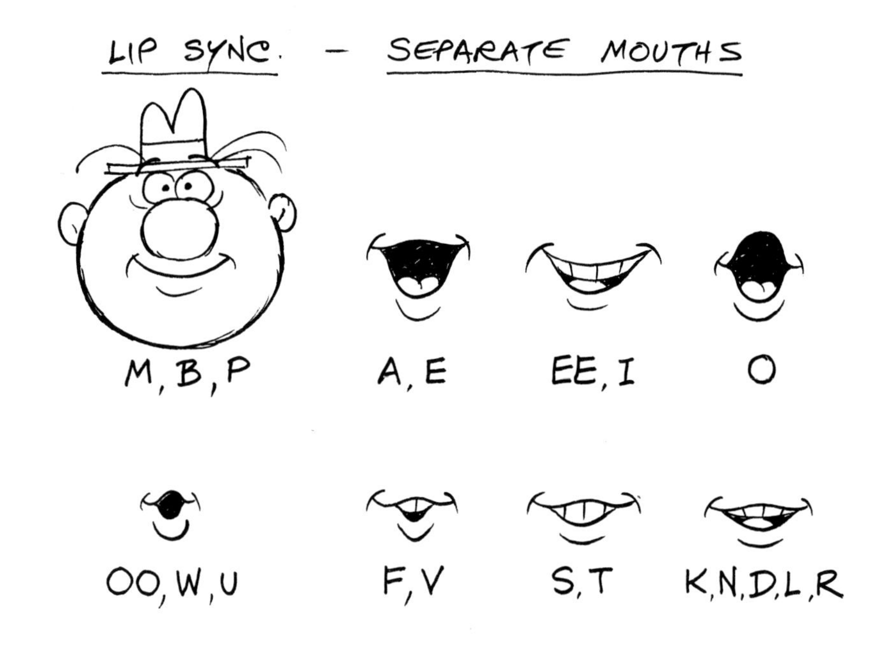

A film with limited animation will almost certainly have fairly full dialogue to keep the plot moving, so mouth movements will need particular attention.

It helps if a character's mouth section has an outline to separate it from the rest of the face – to avoid difficulties in matching. When using separate mouths, about eight or ten mouth positions should be sufficient for each head position. In a close-up dialogue scene it is usually enough to have two poses, or two head positions to point up the main words in the dialogue.

Animation language is an examination of the components of gestures and body movements, in both humans and animals. It is recognized that such non-verbal communication, which constantly takes place, is a highly complex matter, and for a long time it has been the basis of professional acting techniques. Body language can also be useful in animation language provided it is realized that it should only be used as a starting point and that it only works if it is sufficiently exaggerated. The study of gestures, postures, distances between creatures, facial expressions, individual behaviour patterns and body motions can all add up to a better understanding of how non-verbal interrelationships work.

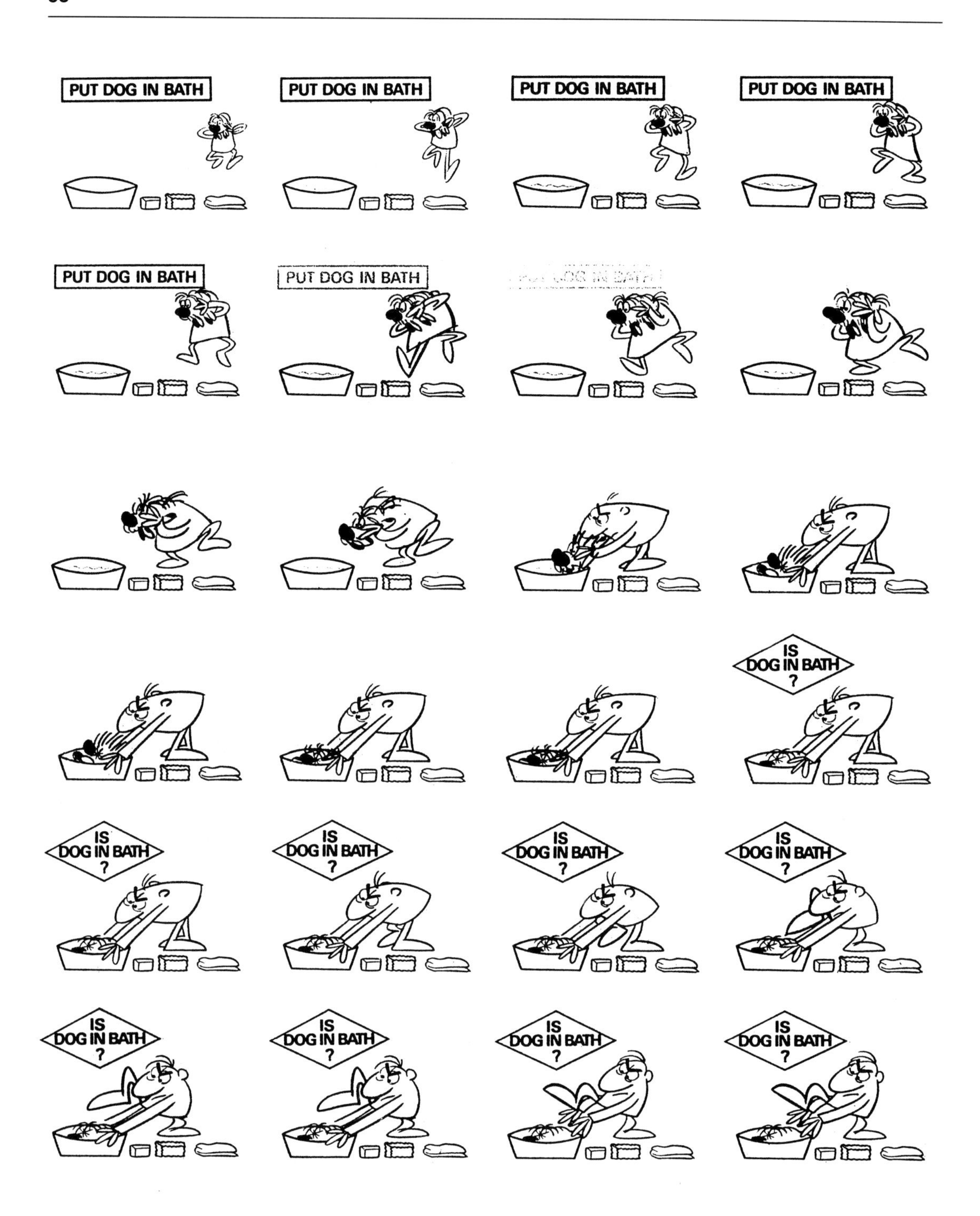

Flow Diagram is an educational film teaching basic computer techniques. The idea is to program washing a dog in a logical sequence. The human character carries out his task eagerly while the dog objects and tries to get away. The situation provides a perfect chance for the characters to express their intention through pantomime body movement

and action. This scene shows how it is possible to economize on animation by using simple outlines and to concentrate only on essential body movement. The animation was done by Harold Whitaker in a combination of single and double frames and produced by the author for British university computer courses.

1 Jealousy
2 Hate
3 Aggression
4 Envy
5 Surprise
6 Determination

Kihachiro Kawamoto is the undisputed master of modern Japanese puppetry. His characters have a wide range of expressions and acting capabilities. A.K. Hering, a former co-worker, writes:

'Japan, of course, holds a significant place in the history of world drama for the quality and depth of its own classical puppet drama, which was gradually elevated to the level of a major theatrical genre from the seventeenth century on. This tradition survives to the present in the incomparable Bunraku puppet drama of Osaka in which each doll is handled by a trio of consummately skilled manipulators, as well as in a variety of other classical or semi-classical forms which still survive as regional arts in Tokyo, on the island of Sado, and in numerous other districts of Japan. Therefore it comes as no surprise to find that Mr Kawamoto has devoted much attention to the technical and visual aspects of this remarkable heritage and that many of his most ultra-modern creative puppets have been expressly constructed for manipulation according to the difficult but singularly effective Bunraku technique.'

6 Soundtrack

A number of animators still wonder whether, when the optical soundtrack was invented in the late 1920s, it was a blessing or a curse for animated cartoons. After all, an animated film can express itself through movement and pantomime and should not necessarily require the help of a soundtrack. *Felix the Cat*, the early *Oswald the Rabbit* and the *Little Nemo* films proved the point, but the soundtrack has been here since 1930, and in fact animated cartoons were the first medium to make the best use of it in several ways:

- In a rhythmical partnership between sound and picture
- In the building of cartoon personalities
- In dialogue and narration
- In sound effects
- In harmonization of sound and vision
- In musical composition
- In a combination of all these

Sound and picture

During the period of reconstruction after World War I, when avant-garde art movements flourished (involving, for example, abstraction and surrealism), it was not surprising that film makers should embrace such newcomers as musical sound and attempt to relate it to picture, and explore its relationship with motion. In the early 1930s experimental artists like Hans Richter and Oscar Fischinger in Berlin were searching for a new unity between the visual and the aural sense, and between mobile graphics and the rhythmical development of musical sound. Among other films, they produced *Brahms' Hungarian Dance* (1931), which brought the two levels of senses as close together as was possible at that time. With this experiment a new type of film making was established, that of abstract film. Both artists devoted all their lives to exploring the inherent potential of visual/music relationships. The field gradually widened and developed into a fundamental art in itself. Others like Len Lye, Norman McLaren and Disney, with *Fantasia* (1941), widened the genre and used it for a variety of entertainment purposes, including pop concerts, classical concerts in theatres and cinemas, musically based graphics for television and advertising, and today computer-generated laser shows.

Cartoon personalities

Practically all popular cartoon characters have been enriched by the voice of a character actor, with the possible exception of Mickey Mouse, whose voice was that of Walt Disney himself. For instance, Mel Blanc did the voices of Bugs Bunny, Porky Pig, Daffy Duck, Tweetie Pie, Sylvester, Speedy Gonzalez and scores of others. Doug Macdonald did Donald Duck, Sterling Holloway's was the unforgettable voice of the stork in *Dumbo* when he delivered baby Dumbo to his mother. Jim Backus did *Mister Magoo*, UPA's myopic genius: without his voice the character would have been meaningless. Specialists like June Foray (Bullwinkle and many others, including some of the latest Disney films) have a peculiar gift for mimicry and voice distortion, which are the basis of characterization for animated characters. The voice artistry can be formidable. When an X-ray was taken of Mel Blanc's throat it was discovered that his vocal chords were similar to those of Enrico Caruso, the great opera singer.

After a careful analysis of each frame and the character of the personalities it is much easier for the animator to maintain established characteristic gestures.

Dialogue and narration

The contribution of a well-written commentary and dialogue can be substantial, making the continuity of a film clear and helping the audience to understand a sequence better. What the animator has to avoid is excessive use of them and to make sure that commentary does not duplicate the action. It must only be used to highlight and clarify an obscure situation. If it succeeds it can save a lot of screen time and add to the interest.

The use of dialogue demands the same discipline. Too much of it can slow down the action and arrest the smooth flow of continuity. Dialogue, as a rule, belongs to television plays. If it is not specially written for animation it could handicap a film rather than help it. Animation demands dialogue containing a sense of humour or a dramatic interest. If both are absent it is better to leave it out and depend on action only.

Sound effects

The use of sound effects in animation is vital to create the illusion the director wants. The question is whether an effect should convey a natural sound or a musical one, or one which is specially created for an action. Natural effects provide a sense of reality, such as a train stopping or pulling away, a car starting up, rain or water pouring down, a horse trotting, or a dog barking. A musical effect could be a train chugging along like the musical train in *Dumbo*, or the painful sound of a character crying for help like Boxer in *Animal Farm*. The justification is functional necessity, or exaggeration to sharpen the meaning of an action. Special sound effects are usually recorded for an action where normal natural effects would sound too flat; ships hooting, trains whistling, a tea kettle talking may come into this category. Specially recorded sound effects are more likely to create a more believable animated environment and have a better chance of fitting to the orchestral soundtrack's synchronization.

Harmonization of sound and vision

Today there is no longer a need to question the effectiveness of aural and visual effects. It has been scientifically tested by a leading commercial agency on behalf of television companies. The finding was that if an audience retains an aural message to the extent of 20%, it retains a visual one to the extent of 30%. But when the two are successfully combined, this rises to 70%. The effectiveness of seeing and hearing has consequently been confirmed. Its utilization is practical in all fields of communication, from advertising to teaching, science and entertainment, but especially in animation which, as a rule, is a medium for experiments in vision, sound and music.

One of the most interesting experiments was carried out by John Whitney Sr in Los Angeles. Digital Harmony is a concept developed by Whitney and may serve as a contemporary example of the genre. The function of Digital Harmony is to make electronically generated music visible through computer graphics. Whitney maintains that – taking rhythm, metre, frequency, tonality and intensity as the parameters of music – the computer can integrate and manipulate image and sound in a way that is as valid for visual as it is for aural perception. His films *Matrix* and *Arabesque* are based on digitally composed images generated with an array of 360 points distributed around a circle timed closely to musical rhythm.

Musical harmonization can greatly help traditional animators with their choreography of action. Chuck Jones utilized such relationships in his productions, from Bugs Bunny to Speedy Gonzalez. Here visual elements are linked with musical instruments: bassoon (a laughing monkey), harp (a nightingale), percussion (Columbus monkey), drum (elephant).

Musical composition

Since both the picture in an animated film and the music have a beginning and an end, both are subject to the progression of time. In this respect there is a natural relationship, a mathematical connection between the two. However, it is not advisable for the composer to tell the same story as the picture. They should have a different function. Music should create the atmosphere and assist the film's dynamics. It should have a unity with the picture and help it with musical ideas. It may not need to follow slavishly the rhythmical development of motion, but if it departs too far from it the picture may look out of sync. A subtle counterpoint helps to make the relationship more interesting, especially during prolonged walking scenes. If there is dialogue and commentary, overlapping should be avoided. It is better to put the music in the background, then bring it to the front as soon as possible. The best solution is to attempt dialogue spoken in musical rhythm, a difficult task for both the composer and the actor, but it has been done successfully several times, as in *The Figurehead* (1953) and *Animal Farm* (1954).

Out of the two options of recording music before or after the production, animators prefer prerecorded music in order to time more precisely the action based on the music. The composer also prefers to record beforehand, to allow greater freedom, not being tied down to every fraction of second. The present situation is a 50/50 draw. In the case of post-recorded music it is customary to have a work guide in the form of a piano track to guide the animator.

With the evolution from optical to magnetic recording one can now obtain a far richer sounding music track, which usually contains the sound effects already composed and mixed onto the final track. This method, which provides a broader range of frequencies, makes the final mixing of the soundtrack simpler. There is no longer a need to mix together a dozen tracks, composed effects or music in dubbing. A few tracks can accommodate the music and effects, the dialogue and commentary; the simpler, the better. The solution will depend on whether the track is to be mono or stereo, but such a choice usually depends on the final objective of the production. The problem, presented over 60 years ago, of the relationship between sound and vision is far from being explored. Today with three-dimensional and stereophonic *musique concrète* and electronic recordings, one is able to create a range of musical sounds never heard before. This has already been achieved by Richard Arnell and David Hewson in animation with the soundtrack of *Dilemma* (1981), *Toulouse-Lautrec* (1984) and *Light of the World* (1989). Their electronic scores created a dramatic high point which lifted the dynamic character of the pictures. If one considers these compositions as a modest beginning, one may reach the conclusion that the sound and vision relationships are only in their infancy.

John Whitney Sr has done a substantial amount of research to bring sound and vision closer. Here he explores a new art field by visualizing the harmonic motion of sound waves through computer animation.

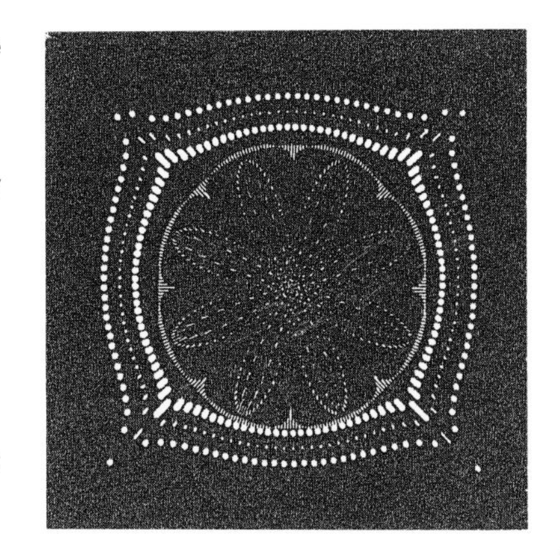

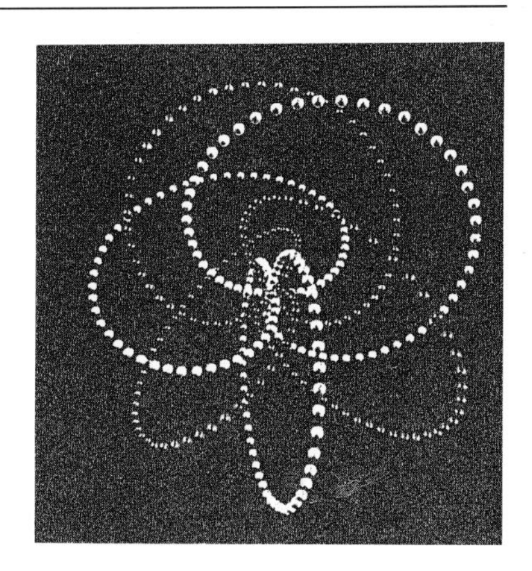

Top: from *Digital Harmony: On the Complementarity of Music and Visual Art.*

Below: sequences from the computer-made film *Pythagoras Revisited*, emphasizing the idea of Differential Dynamics.

From *Digital Harmony* by John Whitney Sr (reproduced by kind permission of John Whitney Sr)

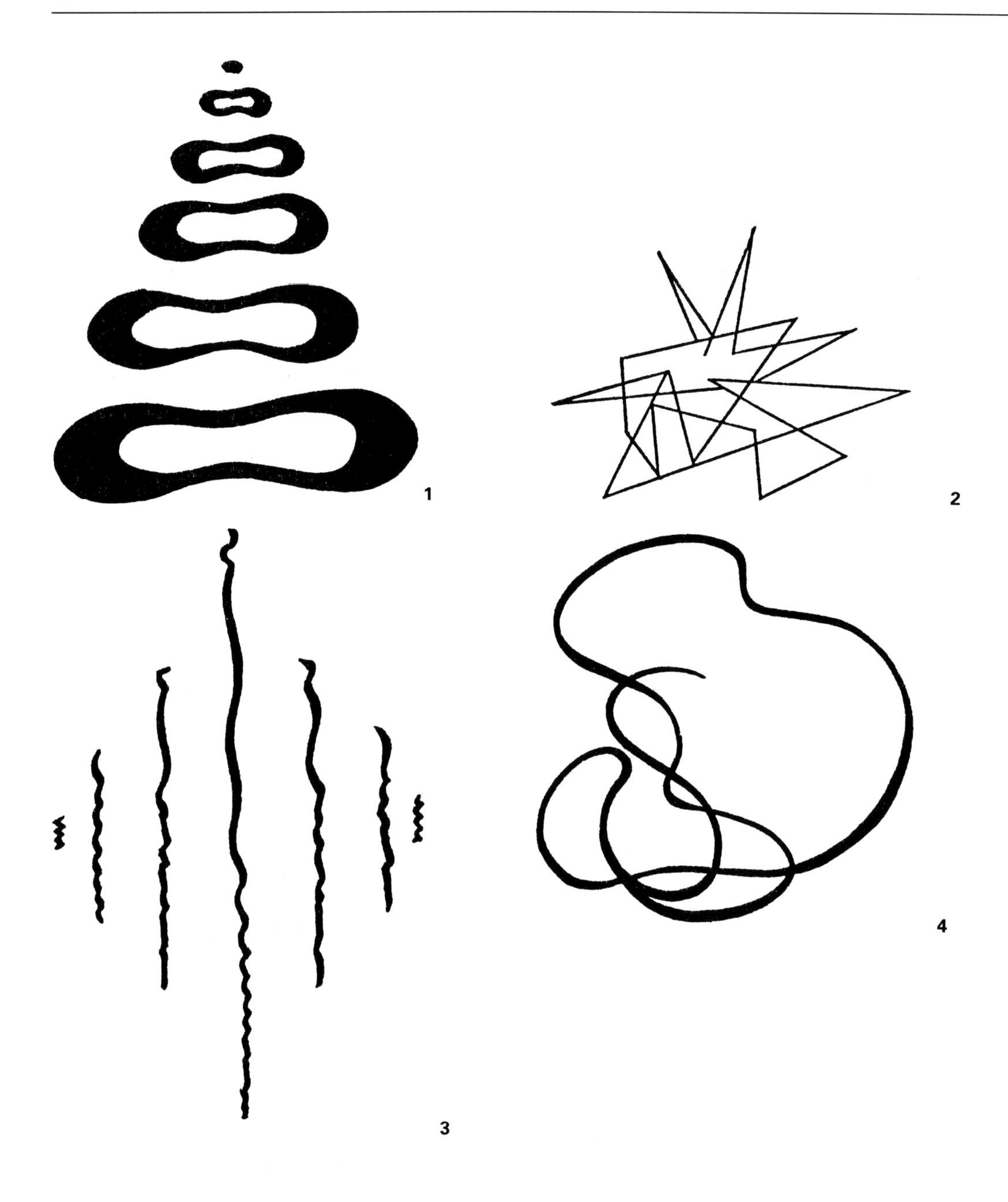

Among the veteran animators of Hollywood, none understands better the strong ties between musical sounds and pictures than Chuck Jones. This instinctive skill has increased the comic power of many characters such as Bugs Bunny, Tom and Jerry, Speedy Gonzalez and Sylvester.

Here he visually symbolizes the sounds of musical instruments:

1 Bassoon
2 Harp
3 Percussion as humming bird
4 Drum as elephant

This sequence by the Swiss animator Rudi Engler depends entirely on fluid animation and its relationship with the sound accompaniment. The music and its rhythmical pulsation is totally integrated with the action.

Rhythmic music: a boy by himself rides a skate-board. A few youths feel challenged by his wish to flee and start to follow him. The chase escalates into a hot pursuit on roller-skates, bicycles, mopeds, fighter planes. While flying in formation above the clouds, however, the protagonists get caught up in another game: the group game. The planes land on the playground and the group continues on roller-skates.

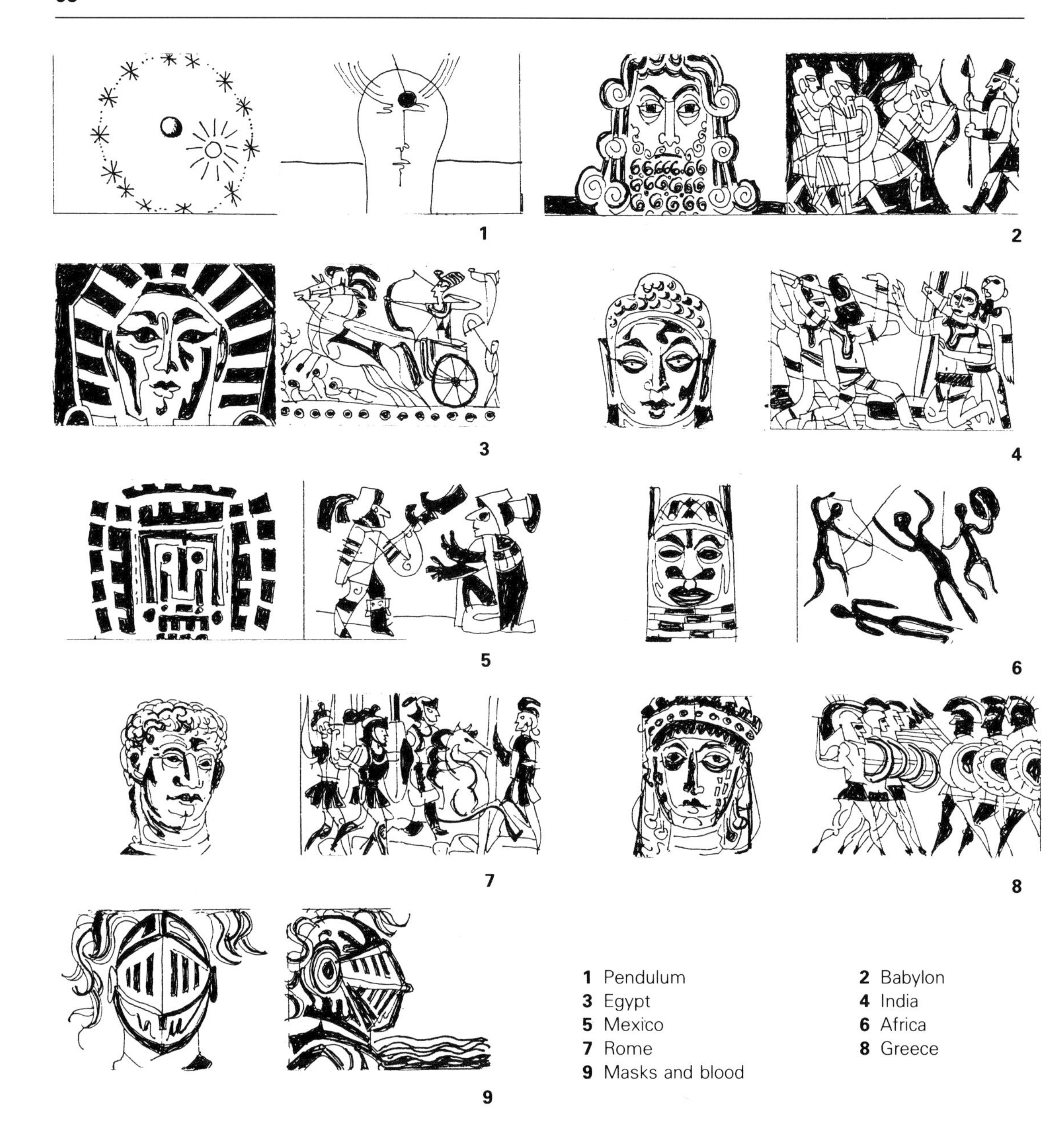

1 Pendulum
2 Babylon
3 Egypt
4 India
5 Mexico
6 Africa
7 Rome
8 Greece
9 Masks and blood

Preplanning the picture and music together is a basic advantage that animation has over live action. In *Dilemma*, a wholly computer-animated film about the development and destruction of civilization, the composers closely followed both the storyboard and the timing of the action. The end result proved that the two levels could be closely united and fit like a glove.

Composers: Richard Arnell and David Hewson
Designer: Janos Kass
Director/Producer: John Halas

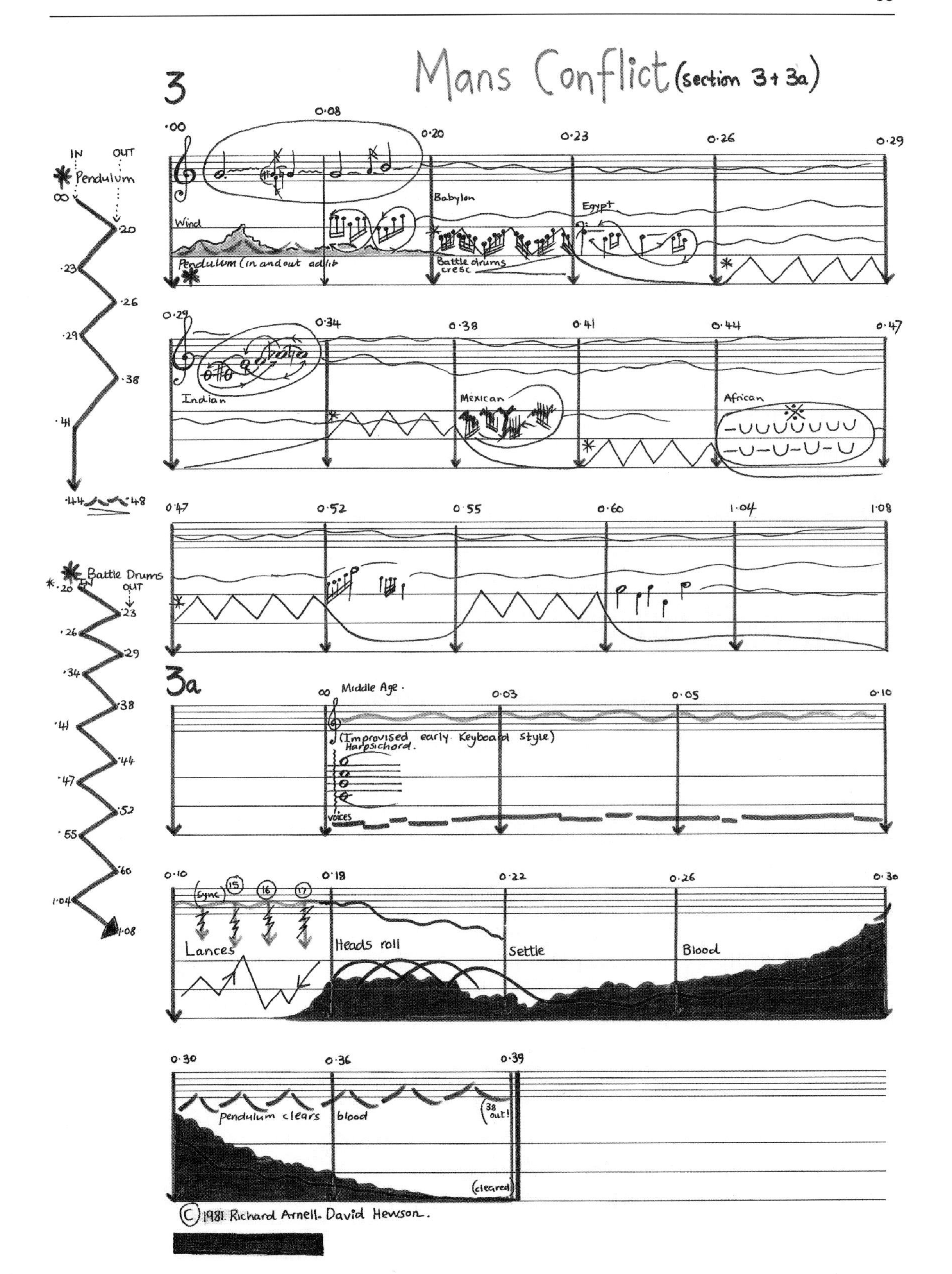

3
Mans Conflict (section 3 + 3a)
·00
0·08
0·20
0·23
0·26
0·29
IN
OUT
Pendulum
Wind
Babylon
Egypt
Pendulum (in and out ad lib
Battle drums
cresc
0·34
0·38
0·41
0·44
0·47
Indian
Mexican
African
0·52
0·55
0·60
1·04
1·08
Battle Drums
3a
Middle Age
0·03
0·05
0·10
(Improvised early Keyboard style)
Harpsichord.
voices
0·18
0·22
0·26
0·30
(sync)
15
16
17
Lances
Heads roll
Settle
Blood
0·36
0·39
pendulum clears blood
38 out!
(cleared)

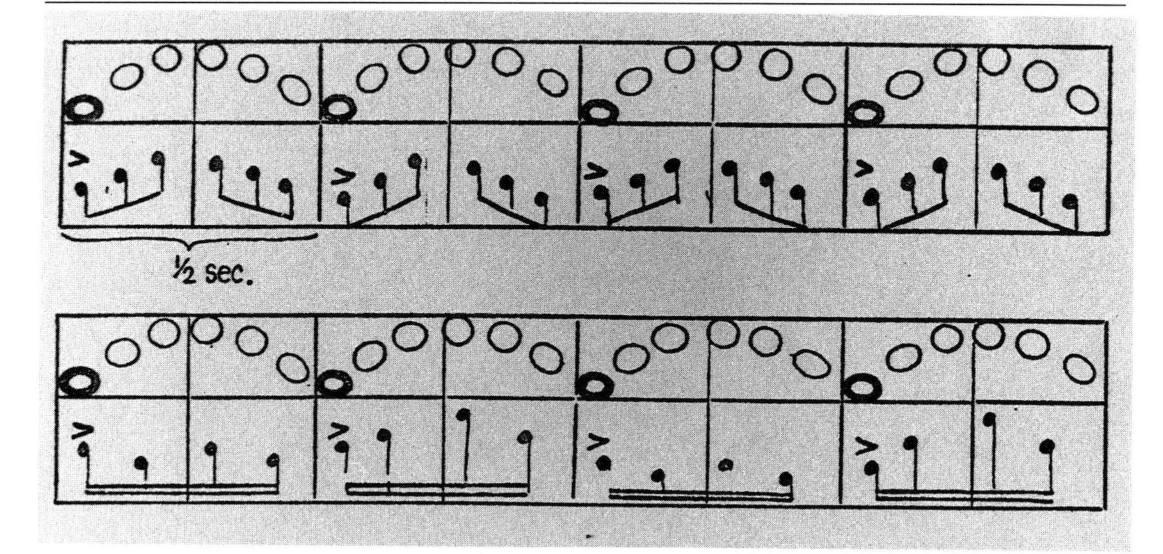

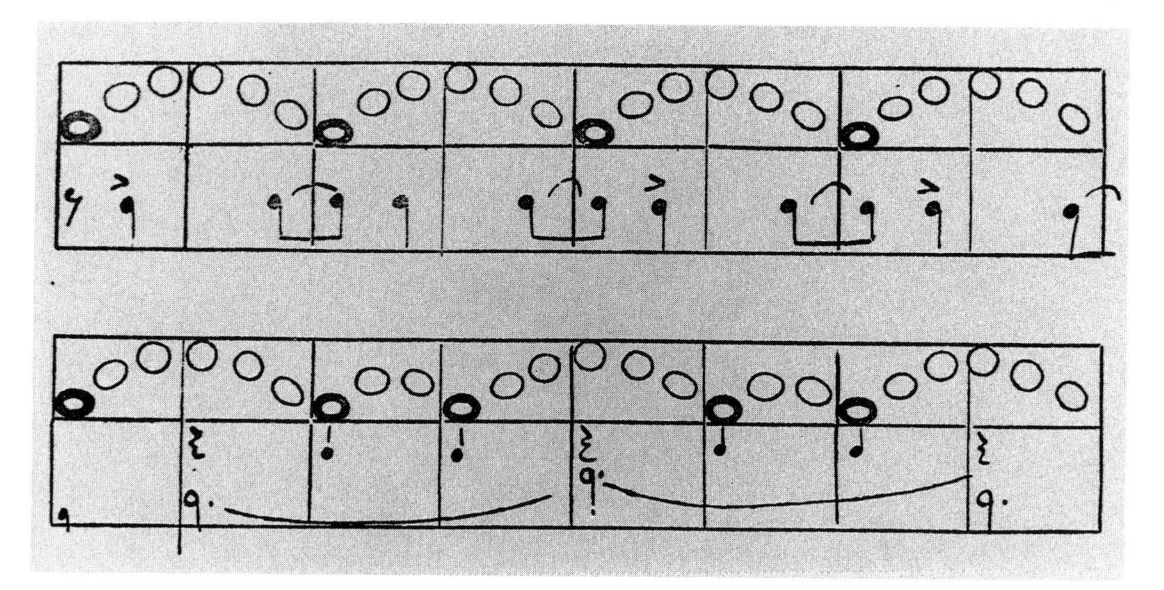

Synchronization between action and music is a constant problem. In these four block diagrams the upper space represents the action, the lower the music. Each block is two seconds long.

1. Action and music are closely related.
2. The music represents mood; it does not slavishly follow the action.
3. The action and musical accent form a close syncopated rhythm.
4. An irregular pattern is stressed by musical rhythm.

Trends change considerably in the relationship of sound and vision. Today, close synchronization, otherwise known as 'mickeymousing' the sound, is discouraged. Instead, one explores the more sophisticated aspect of counterpoint in musical composition as well as the potential of electronic and *musique concrète* sound recordings, which is virtually unlimited.

7 Pictorial survey

Features

With the success of the latest feature-length animated films at the box office, at the beginning of the 1990s the output has increased to 50 features a year, not counting the number of films produced for television. Unfortunately there are not enough skilled animators to cope with the demand, in spite of the gradual utilization of newer techniques. The present list of long films contains a highly interesting range of techniques and styles.

Student films

Since the introduction of animation courses in design and art colleges in the 1970s, the division between professionally made films and those made by students has narrowed considerably. Most students are guided by experienced animators, and if they do not have the most expensive equipment they have basic facilities to practice with. Both experience and practice remain vitally important. These valuable assets cannot be obtained in a short course of two years (or four in continental colleges). A sense of movement and timing can only emerge through years of trial and error, concentration and industry. However, the increased quality of student films is dramatically noticeable in major film festivals, where more and more prizes are awarded to students.

Computer animation

While computer animation contributed substantially to research in science and technology for many years, it did not attract the public's eye until the advertising industry started to use effects like zooms, sparks, motion blurs, self-shadowing and other technical tricks in the 1970s. In the meantime animators learned that computer effects do not necessarily make films which generate emotional contact with the audience. No amount of dazzle can make up for a weak story. Computer animation has to evolve as a tool in the hands of professional animators. Several basic requirements still apply, like story continuity, storyboards, competent movement, and soundtrack. Pages 97–104 present the work of animators and computer technicians who have established a bridge between the two sectors and point towards future development.

International panorama

1 *The Man who Planted Trees,* Frédéric Back, Canada. The film won an Oscar in 1987 and the Grand Prix at Annecy the same year

2 *Slika* (*The Picture*), Rudolf Borošak, Zagreb Film, Yugoslavia

3 *School of Dance,* Peter Szoboszlay, Hungary

4 *Exciting Lovestory,* Borivoj Dovnikovic, Zagreb Film, Yugoslavia

5 *Ogres and Bogies,* Rastko Ciric, Avalfilm Belgrade, Yugoslavia

1

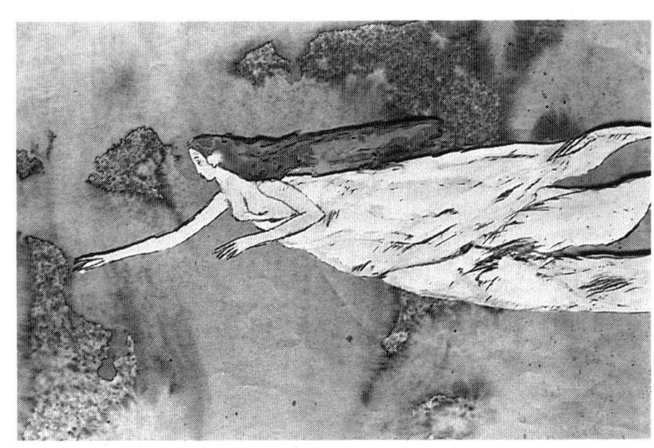

2

3

4

5

1

4

2

5

3

6

1–3 *Butterflies,* Kresimir Zimonic, Zagreb Film, Yugoslavia

4–6 *The Tale of Tales,* Yuri Norstein, USSR

1,2 *If I Were a Bird,* Ondrej Slivka, Czechoslovakia

3 *The Walls,* Piotr Dumala, Poland

4 *The Third,* Homolya Gabor, Hungary

5 Drawing from *A Persian Fairy Tale,* Noureddin Zarrinkelk, Iran

6 *The Fish Dish,* Fang Runnen, China

1

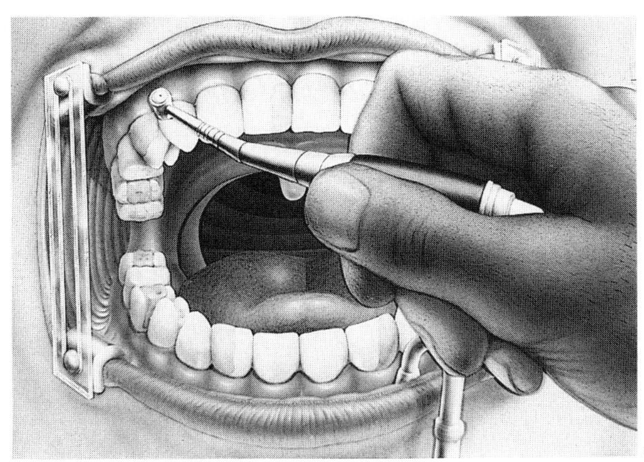

4

2

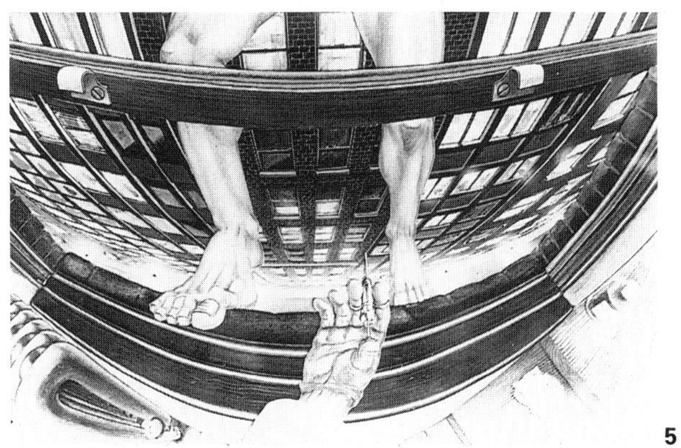

5

3

6

1 *Animated Self-portraits,* David Ehrlich, USA

2 *Logo Interruptus,* Bob Kurz and Friends, USA

3 *Self-portrait,* Renzo Kinoshita, Japan

4 *An Inside Job,* Aidan Hickey, Ireland

5 *Toe Hold,* Aidan Hickey, Ireland

6 *Il Tempo,* a TV opening title, Manfredo Manfredi, Italy

1 *Pas à deux,* Monique Renault and Gerrit Van Dijk, Belgium

2 *Pictures from Memory,* Nedeljko Dragic, Zagreb Film, Yugoslavia

3,4 *Only a Kiss,* Guido Manuli, Italy

5 *Ubu,* Geoff Dunbar, Great Britain

6 *Pepere et Memere,* Federico Vitali, Italy

1

2

3

4

5

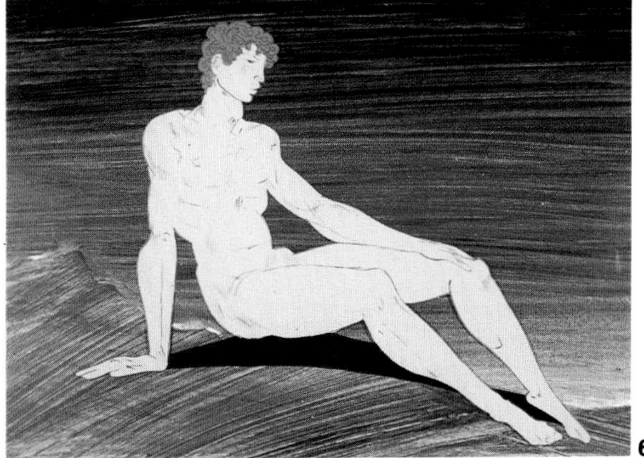
6

7

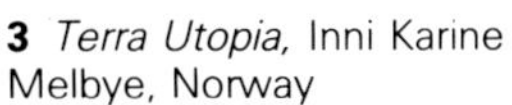

1 *Pulcinella,* Gianini and Luzzati, Italy

2 *La Belle France,* Bob Godfrey, Great Britain. Made for the Revolution anniversary in 1989

3 *Terra Utopia,* Inni Karine Melbye, Norway

4 *Farewell Little Island,* Sandor Reisenbuchler, Hungary. Grand Prix, Kecskemet, 1988

5 *Every Child,* Eugene Fedorenko and Derek Lamb, Canada. The film was produced for the UNICEF Year of the Child in 1979 and won an Oscar

6 *The Light of the World,* John Halas, Great Britain

7 *Paradisia,* Marcy Page, USA

1

4

2

5

3

6

1 *Freckle Juice,* Barrie Nelson/Judy Blume, USA

2 *Baeus,* Bruno Bozzetto, Italy

3 *The Discovery,* from *'The Bee and the Frog',* Gunter Rätz, DDR

4 *How to Kiss,* Bill Plympton, USA

5 *25 Ways to Quit Smoking,* Bill Plympton, USA

6 *Fantabiblical,* Guido Manuli, Italy

1 *The Wind,* Csaba Varga, Hungary

2,4,5 *City,* Rein Raamat, Tallinfilmstudio, Estonia

3 *Waltz,* Csaba Varga, Hungary

1

2

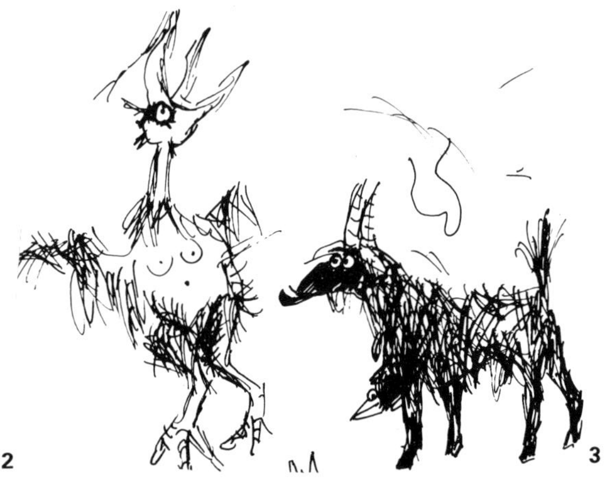

3

4

5

1

2

3

Student films

1–3 *The Hill Farm,* Mark Baker, National Film and Television School, Great Britain. Won Grand Prix, Annecy, 1989. Mark Baker took three years to make his film, influenced by the style of English painters; he used multiple exposures for the final effects

4 *Trixtown Tribulation,* Steve Arnott, Royal College of Art, Great Britain. Steve Arnott won the Student Animator of the Year Award from the British Animation Awards, 1988

4

1

4

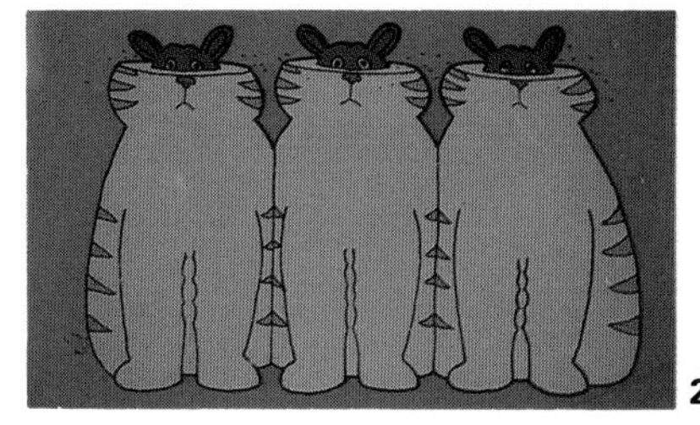

2

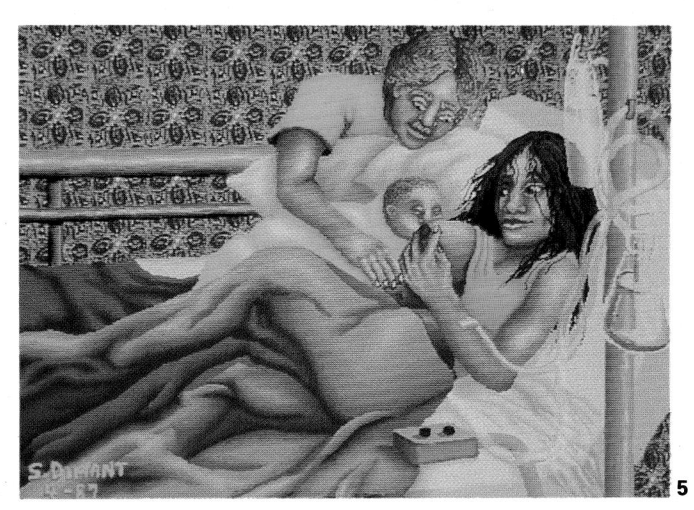

3

5

6

1 *Hello Dad,* Christoph Simon, Royal College of Art, Great Britain

2 *Jollity Farm,* David Stone, Royal College of Art, Great Britain

3 *Madame Potatoe,* Emma Calder, Royal College of Art, Great Britain

4 *Grand National,* Susan Loughlin, National Film and Television School, Great Britain. The technique used was animating directly with brush and ink onto paper

5 *Birth,* Suzanne Dimant, CalArts, California

6 *The Spiritual in American Popular Culture Series,* Bill Luttrell, CalArts, California

1 *Under the Same Sun,* Nick Hellman, West Surrey College of Art and Design, Farnham, Great Britain

2 *City Number Seven,* Suzane Mitus

3 *Polo Rider,* Mike Merell, CalArts, California

4 *The Last Post,* Andy Wyatt, Harrow College of Art, Great Britain. The film combines images drawn directly onto film with live archive footage

1

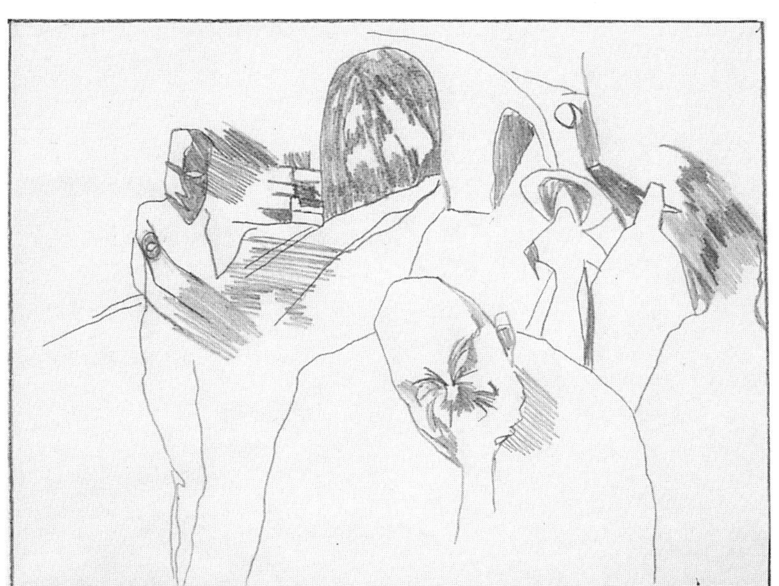

2

3

4

1 *Dreamchild,* Lyndon Gaul, Bournemouth and Poole College of Art and Design, Great Britain. The animation was drawn in the traditional way and then reworked and line tested until correct. The drawings were then taken in to the Quantel Paintbox and artworked on it. Much of the light and detailed animation was improvised

2 *Found Guilty,* Tobias Fouracre, West Surrey College of Art and Design, Farnham, Great Britain

3 *Still Life,* Yaki Kaufman, Bezalel Academy of Art and Design, Jerusalem, Israel

4 *Survival,* Mona Abo El Nast, CalArts, California, under the guidance of Jules Engel. The film is influenced by traditional forms of Egyptian art

1

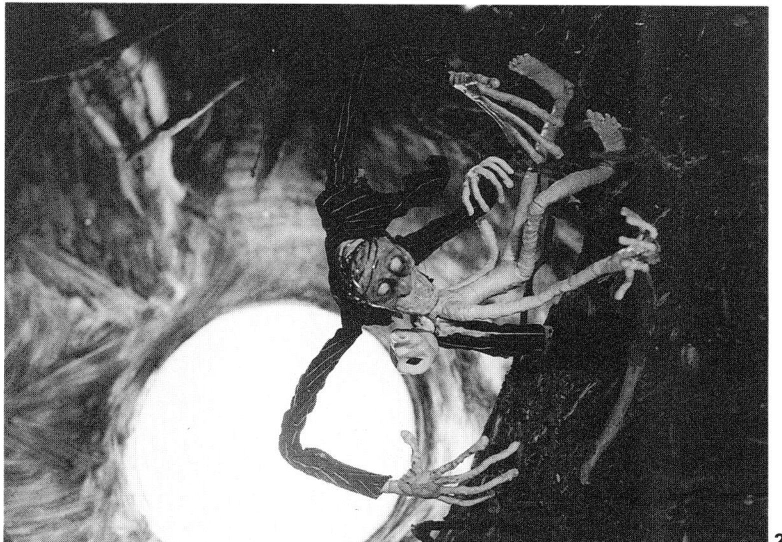

2

3

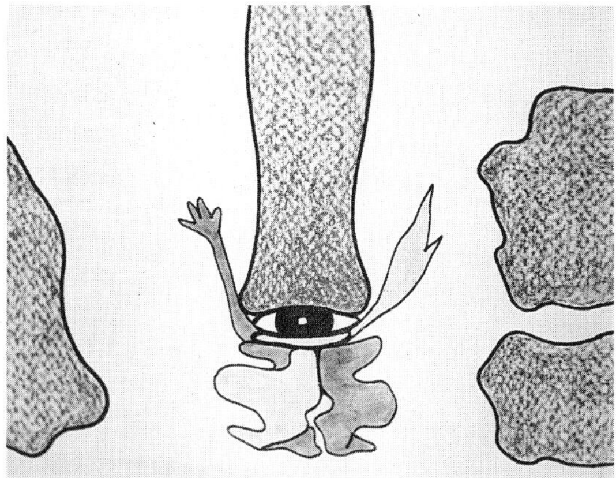

4

Feature films

1–3 *Oliver and Company,* Walt Disney Productions, USA

1

2

3

1–6 *Asterix in Britain,* Pino van Laamsweerde, France

1

2

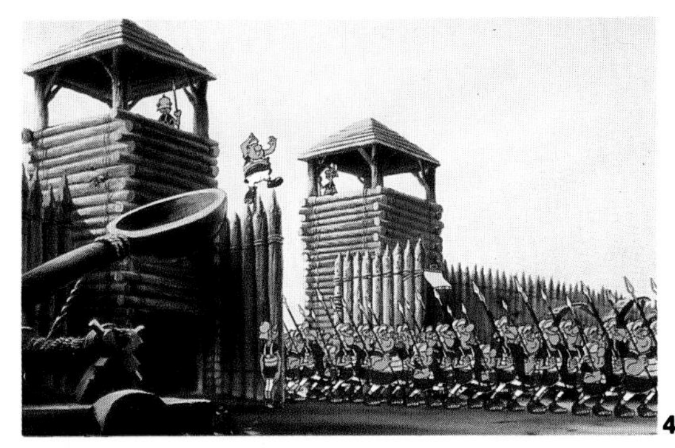

4

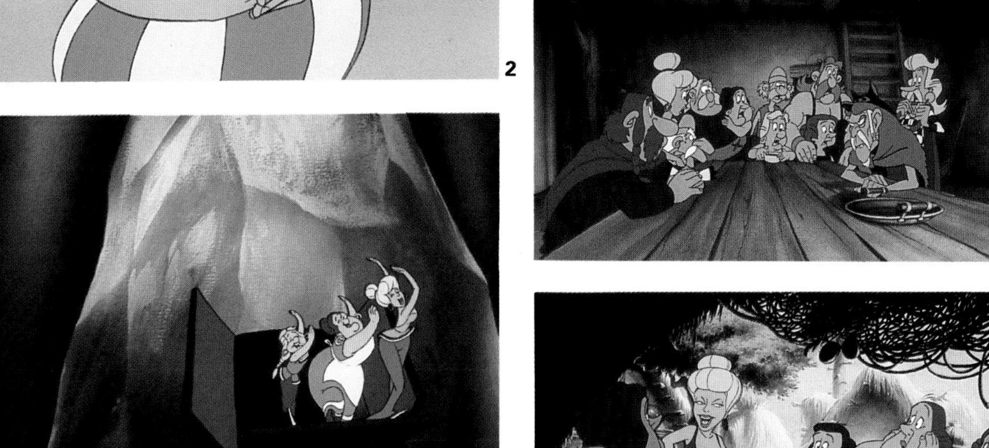

5

3

6

1

2

3

1 *Gwen,* Jean-François Laguionie, France

2 *Gandahar, René Laloux, France*

3 *Heroic Times,* Jozsef Gemes, Hungary

1

2

1 *Pilgrim's Progress,* John Halas, design Janos Kass, Great Britain

2 *The Adventures of Mark Twain,* Will Vinton, USA

3 *When the Wind Blows,* Jimmy Murakami, Great Britain

4 *Fables of La Fontaine,* Attila Dargay, Hungary

3

4

1–3 *Valhalla,* Swan Film Production, Denmark

4 *Son of the White Mare,* Marcell Jankovics, Hungary

5 *The Quest of Sindbad,* Noureddin Zarrinkelk, Iran

1

2

3

4

5

1–3 *Taxandria,* Raoul Servais, Belgium

1

2

3

1

Object animation

1 *A Crushed World,* Boyko Kanev, Bulgaria. The film was made using hardened newspaper material. The handling of it is convincing. The film won the joint Grand Prix at Annecy, 1987

2, 4 *Dreamless Sleep,* David Anderson, Great Britain

3 *The Warrior,* Jin Xi, China. Jin Xi is able to produce soft motion out of hard wood puppets. He controls motions of head and limbs with ball joints. He paints the features, and decorates his figures following the grain of the wood.

2

3

4

1 *The Great Cognito,* Will Vinton, USA. Will Vinton has developed a great deal of artistry in the handling of clay material. With *The Great Cognito* he handles it with a masterly assurance, and shows a sense of humour no-one assumed to be possible with clay

2 Lurpak *Hang Glider* commercial, Peter Lord and David Sproxton, Aardman Animations, Great Britain

3 *Babylon,* Peter Lord and David Sproxton, Aardman Animations, Great Britain

4 *The Adventures of Mark Twain,* Will Vinton, USA

1

2

3

4

1 *Augusta Feeds her Baby,* Csaba Varga, Hungary. The rapid rise of Csaba Varga is due to his unique sense of timing in three dimensions, using solid materials like wood and softer ones like clay and cloth. He adjusts his treatment according to the material, making the best of its nature.

2, 3 *Cupboard Tales,* Csaba Varga, Hungary

4 *Speed Demon,* Will Vinton, USA

1–3 *To Shoot Without Shooting,* Kihachiro Kawamoto, Japan/China. Kawamoto, who studied under Jiri Trnka in Czechoslovakia, retains an oriental influence in his puppet films

4 *Treasure of the Grotoceans,* Co Hoedman, Canada

1

2

3

4

1

2

Experiments

1 *Dissipative Dialogue,* David Ehrlich, USA

2 *Chips,* Jerzy Kucia, Poland. Jerzy Kucia with his impressionistic textural animation helped to advance visual poetry on the screen

3 *The Overcoat,* Yuri Norstein, USSR. This outstanding artist constantly comes up with novel solutions not yet explored in animated films

4 *Omad Povrod,* Rein Raamat, Estonia

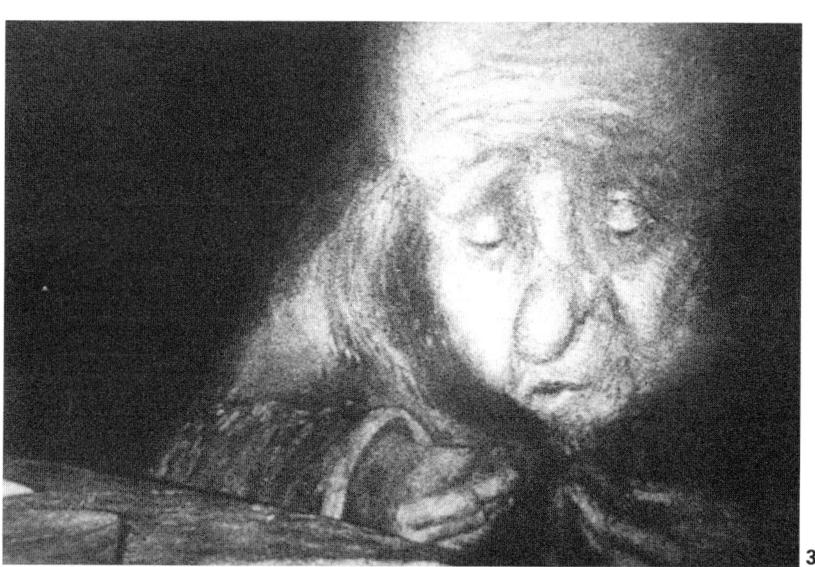

3

4

1

2

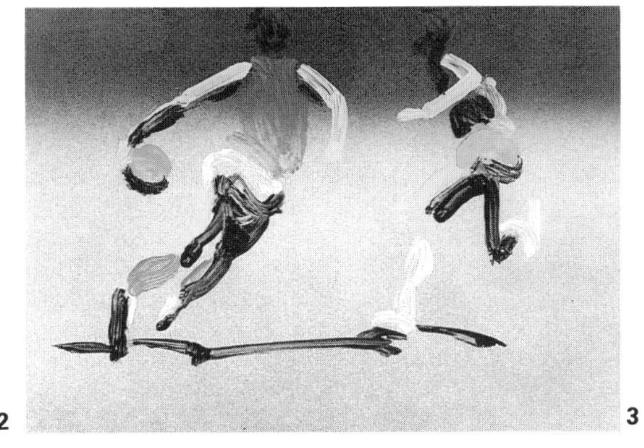

3

1 *Manly Games,* Jan Svankmajer, Czechoslovakia

2, 3 *Hors-Jeu,* Georges Schwizgebel, Switzerland. A combination of rhythmic animation with spontaneity of free brushstrokes typifies the work of Schwizgebel

4 *Terminal Image,* Elrik, France

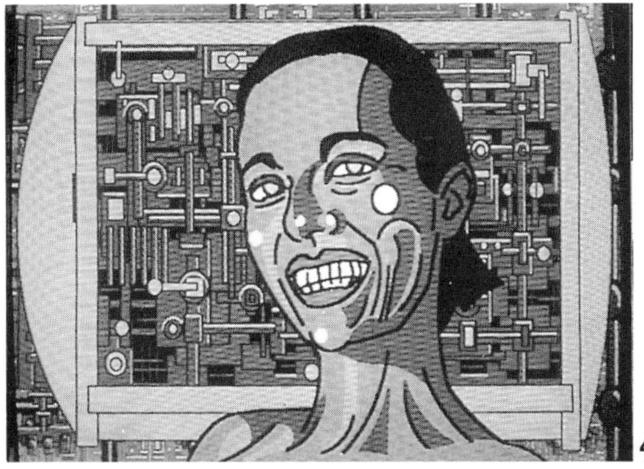

4

1

1 *O Atomo Brincalhao,* Roberto Miller, Brazil. Intuitive brush arts brought onto optical celluloid

2 *Ab Ovo,* Ferenc Cako, Hungary. Animated entirely with shifting sand, a material extremely difficult to control

3 *Interior,* Jules Engel, CalArts, USA

4 *Landscape,* John Adamczyk, CalArts, USA

2

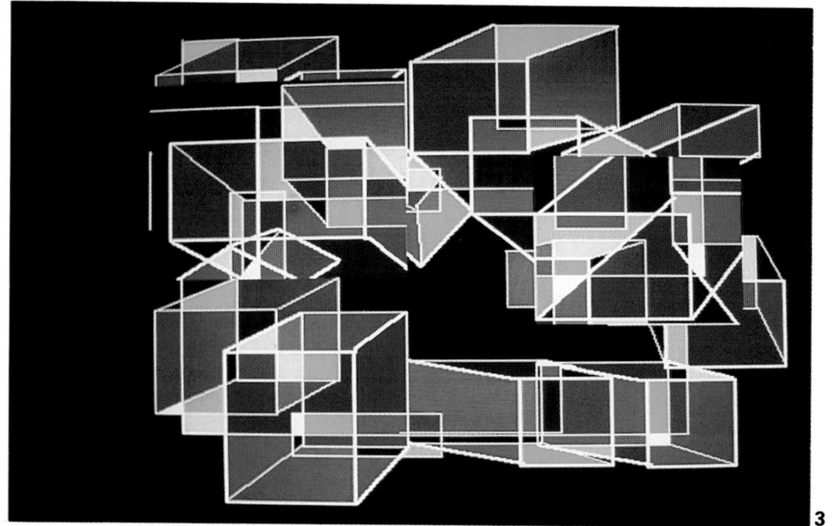

3

4

Computer animation

1 John Lasseter has made a number of memorable animated shorts at Pixar, San Rafael, California, among them *Luxo Jr* and *Red's Dream*. *Tin Toy*, winner of the Academy Award in 1989, was produced by Pixar's Animation Production Group. Direction, animation, story: John Lasseter. Technical direction: William Reeves, Eben Ostby. Additional animation: William Reeves, Eben Ostby, Craig Good. Modelling: William Reeves, Eben Ostby, John Lasseter, Craig Good.

2 *Peppy*, Art Durinski and Michiko Suzuki, Links Corporation, Tokyo. Hardware: Links-1 system, VAX-II/780. Software: Ray Tracing (Tracy), Scanline

3 *Alphabet 'K'*, Links Corporation, Tokyo. Designer: Taku Kimura. Hardware: Links-1. Software: Ray Tracing

4 *Intelligent Service*, Tokyo Gas TV commercial for Hakuhodo Inc and Caravan Inc. Producer: Kinnosuke Morikawa (Caravan Inc), Kinji Odaka/Shuji Asano (Links Corp). Director: Jun Asakawa (Caravan Inc), Hiroyuki Hayashi/Midori Yamada (Links Corp). Hardware: Vax 780. Software: in-house Scanline software

1

2

3

4

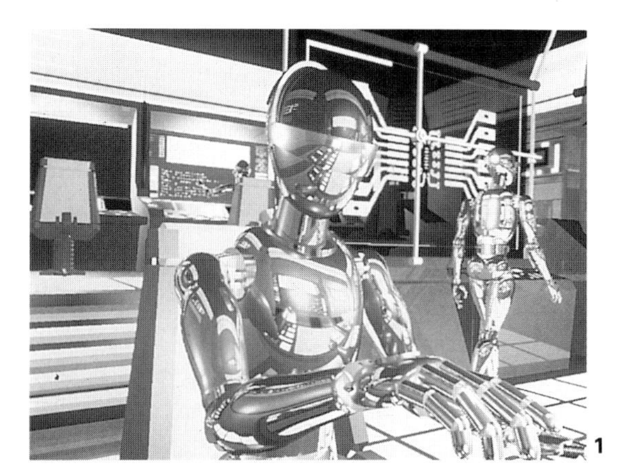
1

2

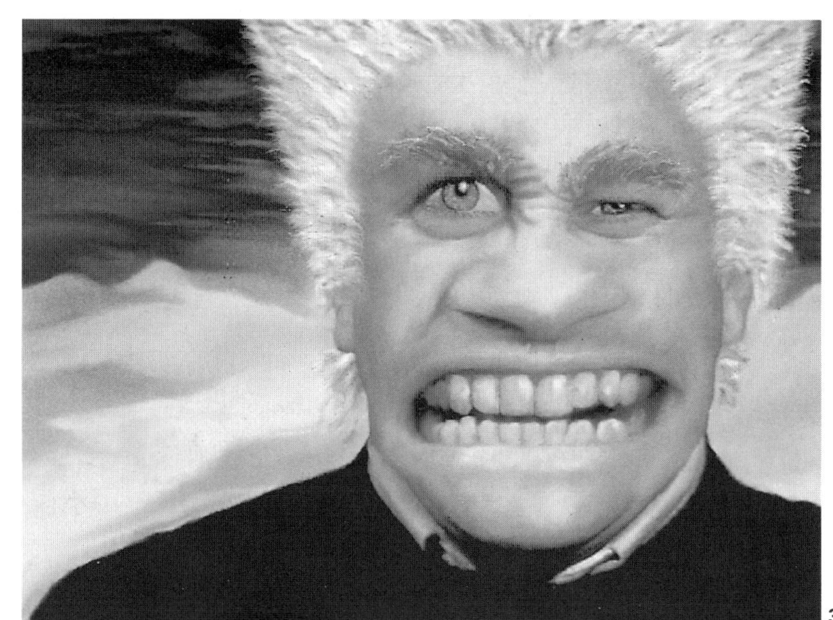
3

1 *Kinsei,* TV commercial for Dentsu Incorporated. Producer: Jun Ueno (Dentsu Inc), Shuji Asano (Links Corp). Director: Taku Kimura (Links Corp). Hardware: Links-1. Software: Ray Tracing (Tracy)

2 *In My Work,* Links Corporation, Tokyo. Designer: Hiroyuki Hayashi. Hardware: Links-1. Software: Ray Tracing

3 *Big Peter*
4 *John Three*
5 *Tom*
1 (page 99) *Eyes*
Peter Voci, Media and Art Center, NYIT.

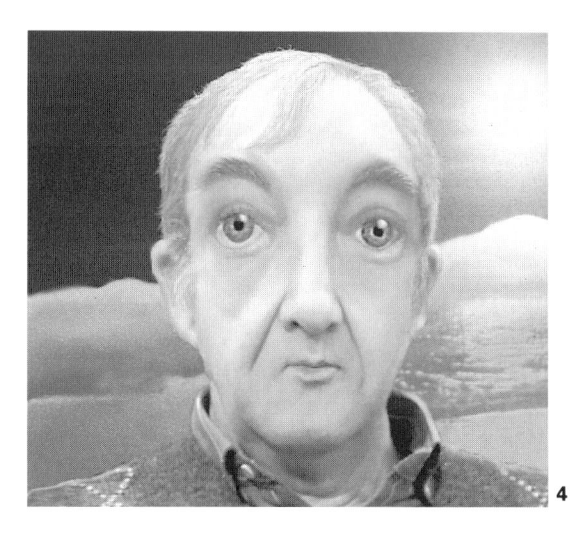
4

5

Peter Voci, Associate Professor and Computer Graphics Co-ordinator at New York Institute of Technology, provides the following technical description: 'The screen images are created utilizing a paint system running CGL's Images II+ software on a DEC Micro PDP 11. The Synthesis Portraits are electronically painted in 2000-line resolution and sectioned in a four by four field arrangement. Each section is then stored on disk as a picture file and then output to various Ink-Jet Printers. The same files are also translated into another format and are output to other devices.

2, 3 Cranston-Csuri unit (see 1–4 on page 100)

4 *Kheops – Head of Arab,* V. Sokolov, programme ALEX, for Display Station Gamma 4:2, Moscow

5 John Halas, Great Britain. A Hieronymus Bosch character for *Inferno*, a computer film processed through Alan Kitching's ANTICS system

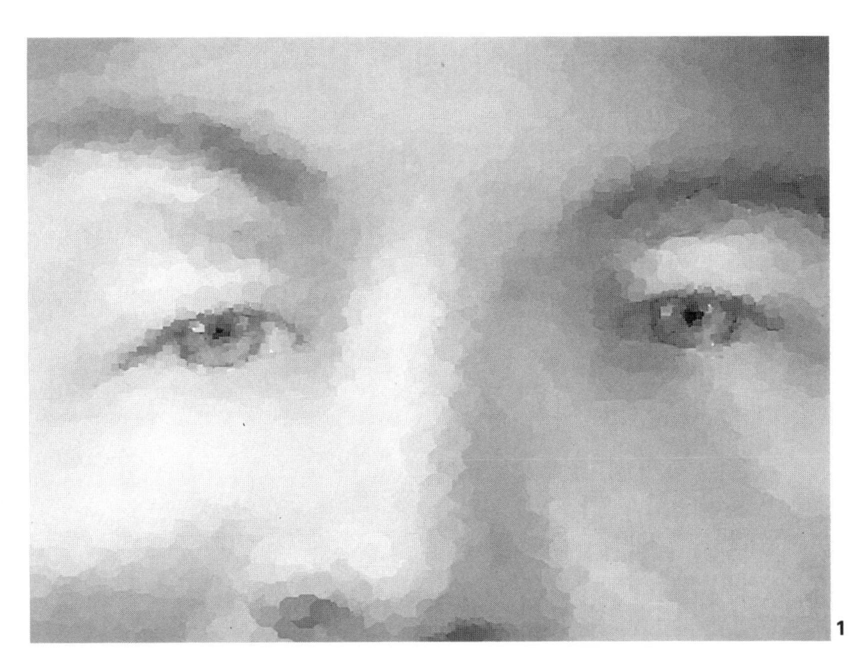

1

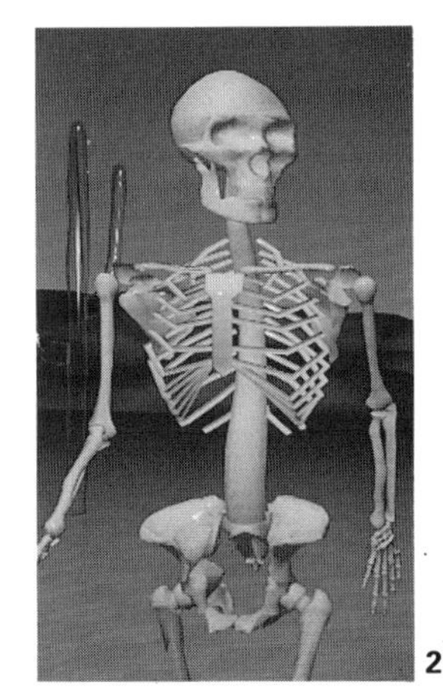

2

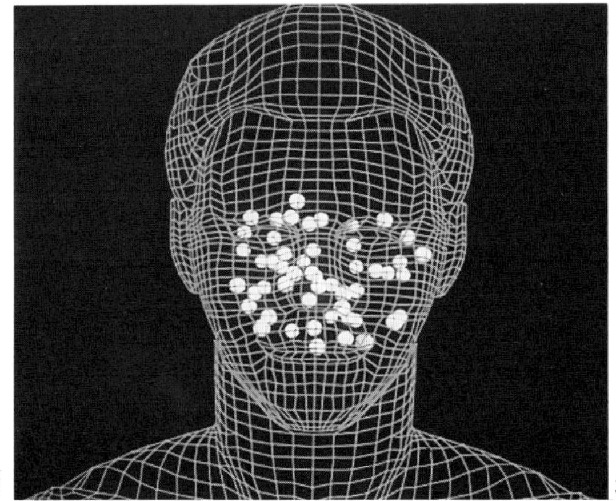

3

4

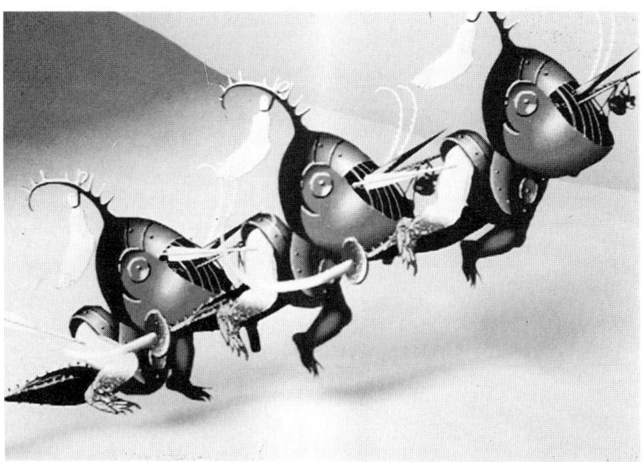

5

1

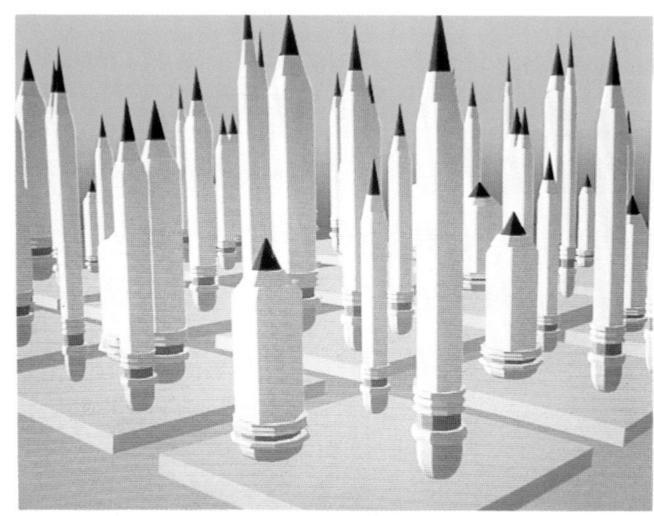

2

3

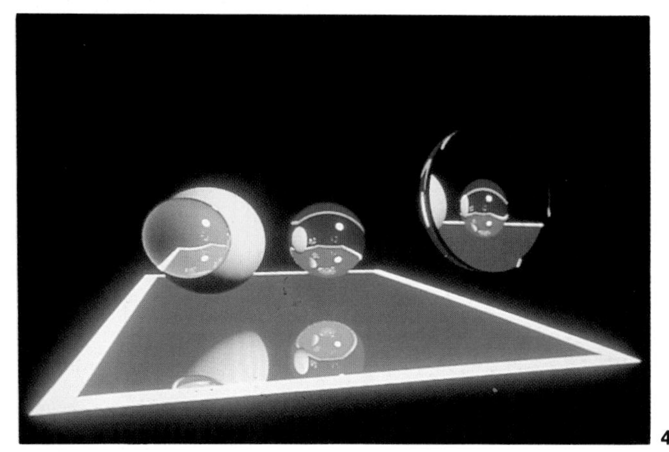

4

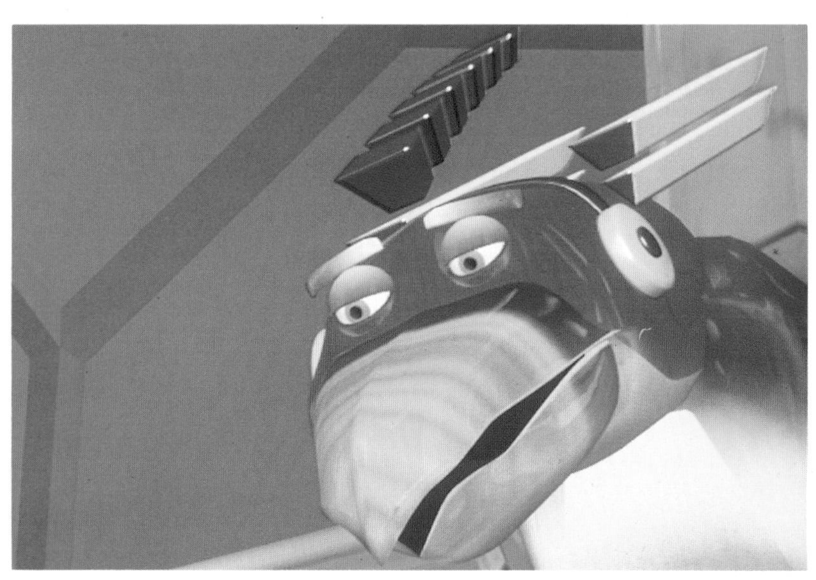

5

1–4 Under the technical guidance of Professor Charles Csuri, the unit of Cranston-Csuri in Columbus, Ohio, has an outstanding record in exploring art and technology

5 Philippe Bergeron established himself through the National Film Board of Canada's film *Stanley and Stella: Breaking the Ice* in the Symbolic Graphic Division with the Whitney-Demos unit, USA, in 1988. It was presented by INA at the Monte Carlo Festival in 1988

1, 2 Hervé Fischer's *Le Chant des Etoiles*, directed by Michel Meyer of France, a 12-minute space venture with a cosmic orchestra. Co-production between La Cité des Arts et Nouvelles Technologies de Montréal (Quebec) and INA in Paris. Modelling software and animation by TDI on IRIS 3020 graphic station by Silicon Graphics. Calculations by CONVEX. First prize at NCGA, USA, in 1988

3–5 Ron Feraglio at Sonic Vision, Australia, uses Bosch's FGS 4500 for 3D animation

1

2

3

4

5

1

2

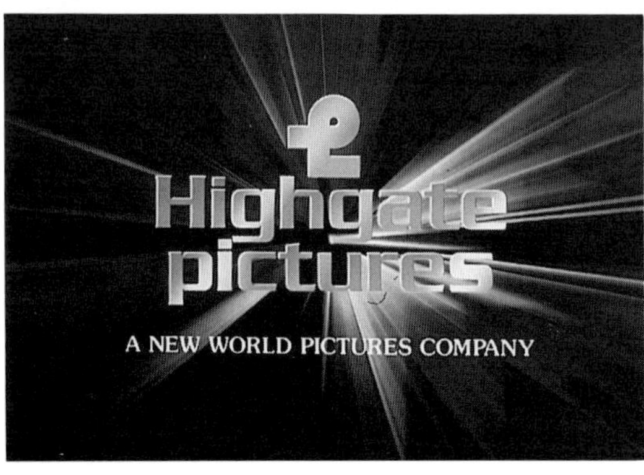

3

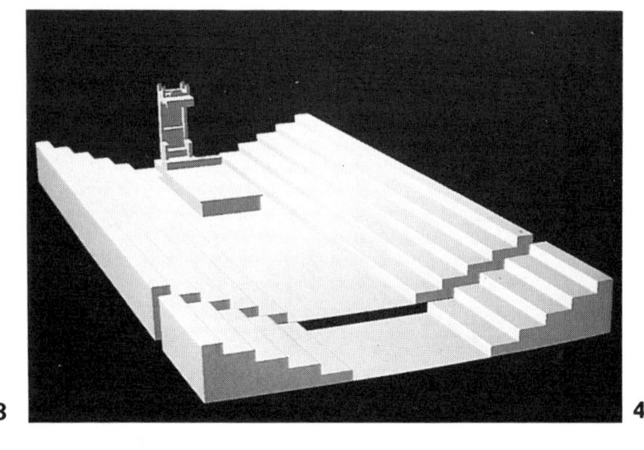

4

1 *CBS Late Night,* Pacific Data Images, Sunnyvale, California, USA. Creative director: John LeProvost, CBS. Associate creative director: Lewis Hall, CBS. Animator: Roger L. Gould, PDI

2 *CBS Evening News with Dan Rather Promo,* Pacific Data Images, Sunnyvale, California, USA. Produced for California Film. Creative director: Wendy Vanguard. Animator: Adam B. Chin, PDI

3 Judson Rosebush of Rosebush Vision Corporation, New York, is one of the early pioneers of digital animation in the USA

4 *Election in Great Britain,* Chris Fynes and John Spiers, Crown Computer Graphics, London, for BBC TV

1, 2 (page 103) Animatica in Barcelona, Spain, a new unit, led by M. Xavier Berenquer, equipped with Bosch FGS 4500 and an IRIS 3020, is one of the busiest studios in Europe

1

2

3 *Eurythmy,* Amkraut and Girard, Ohio State University

4, 5 *Dirty Power,* Robert Lurye, Ohio State University, USA. Robert Lurye writes: 'This project has been partly an experiment with joint interpolation using an Evens and Sutherland PS300 and "twixt" animation software to create anthropomorphic plugs. It has also been an experiment with a new image renderer which has enabled me to create interesting stylized lighting effects. The images for *Dirty Power* are being created using a network of SUN microsystems computers and a Convex C1 computer.

3

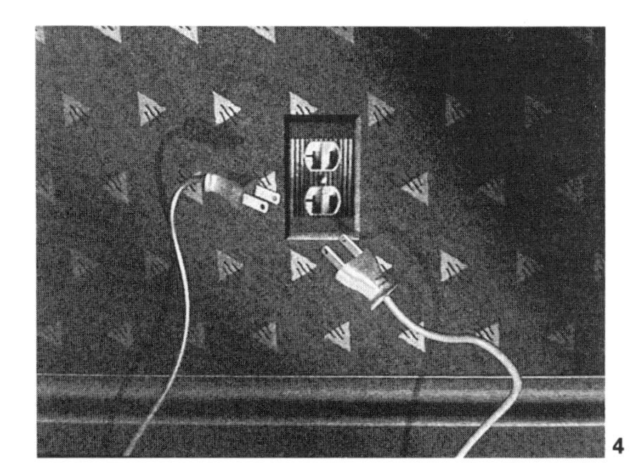

4

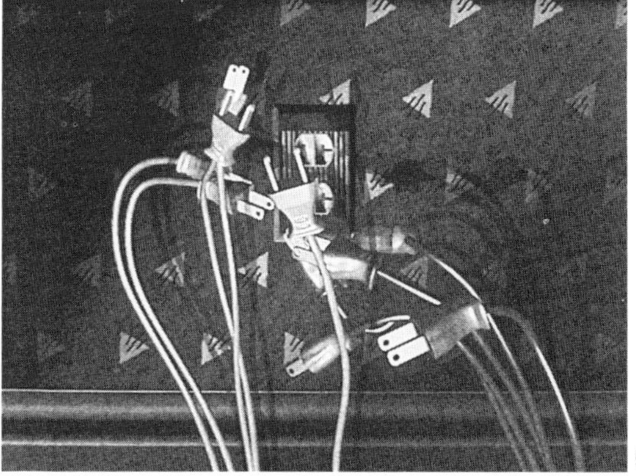

5

1

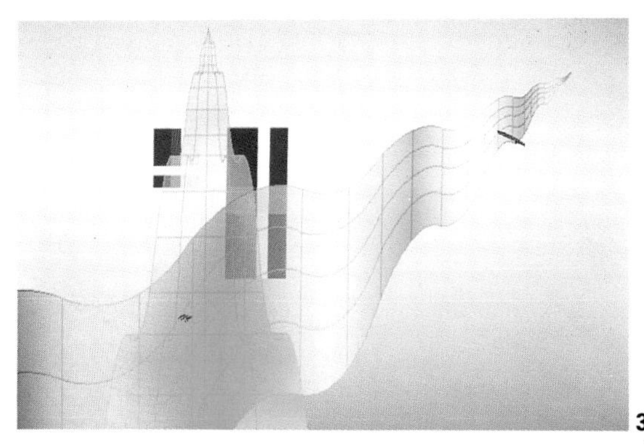

3

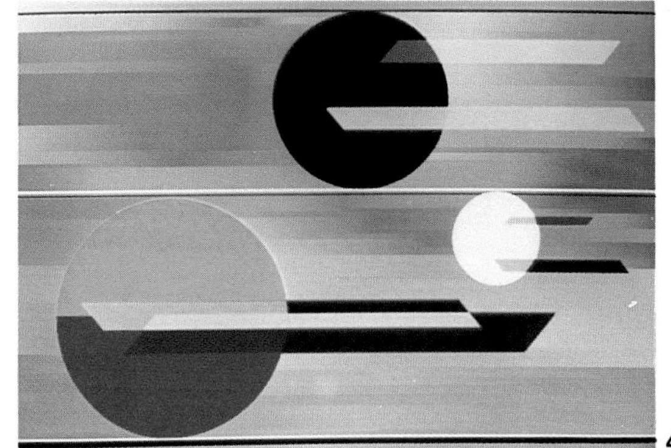

4

2

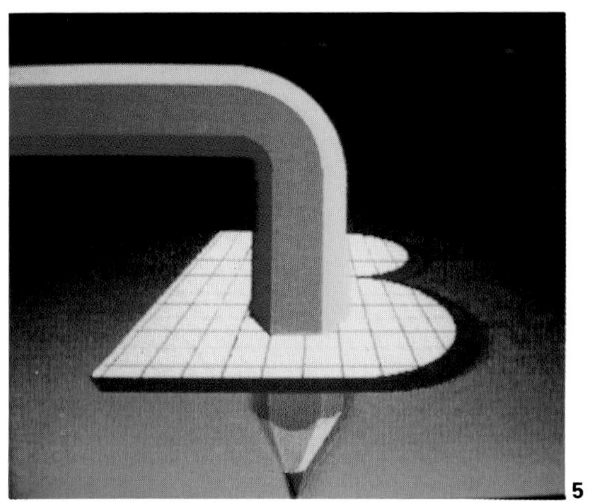

5

1, 2 Jim Lindner's Fantastic Animation Mechine Inc in New York is a comparative newcomer. They have already made a strong impact with their well-equipped facilities, to serve clients like IBM, Fuji, Timex and Monsanto

3, 4 John Halas and Tamas Waliczky, Great Britain/Hungary. From a film tribute to the pioneering work of Moholy-Nagy

5 *Riga*, Victor Zakiev, artist of the TV Company Ekran Central TV, Moscow, from IBM-AT with the frame buffer Supernova

Glossary of animation terminology

Traditional

A & B roll printing A method of introducing opticals in a film without resorting to duplicate negatives. The negative of the film is assembled in two rolls, a scene that requires mixing out being on one roll, the next scene that requires mixing in on the other. The two fades that constitute the mix are effected by a shutter mechanism on the printer.

Academy gate The standard gate used for 35 mm film except when wide-screen processes such as Cinemascope are being used. The dimensions of 35 mm Academy and standard 16 mm camera gate apertures are as follows:

Standard		
Width	0.404 in	10.26 mm
Height	0.295 in	7.49 mm
35 mm Academy		
Width	0.864 in	21.95 mm
Height	0.63 in	16.0 mm

Academy leader Head and tail leaders of release prints, designed by the Academy of Motion Picture Arts and Sciences. Standard throughout the motion picture industry.

action A term used to differentiate between the picture and sound of a film, action being the picture, sound being the sound. *Mute* is a term used to mean the picture film or negative as opposed to the sound film (optical or magnetic).

aerial image A technique which combines the images of processed film with those on cel by photographing them together. A projector is focused at the level of the compound tabletop. When cel images are interposed in the path of the projected image, a combined image is presented to the lens of the camera and is photographed.

anamorphic lens A lens which squeezes the picture image in a camera or unsqueezes it in a projector. The squeeze is normally horizontal and is used in such processes as Cinemascope.

animated insert An insert is a shot, often a close-up, cut into a sequence and often shot separately from the rest of the sequence. If the shot is made by the animation process it becomes an 'animated insert'. See also *sequence*.

animation The process of producing an illusion of motion in a film or video when no motion has in fact taken place. Thus animated cartoons, animated diagrams and animated puppets are all examples of animation but time-lapse photography, slow-motion or string puppets (marionettes) are not. The art of animation is the creation of moving images through the manipulation of all varieties of techniques apart from live-action methods.

animation cycle Often in animation a cycle of cels can be repeated again and again in the same order. For example, a character walking against the panning background might contain ten or a dozen cels which can be repeated for the whole length of the shot. Such cels are known as an 'animation cycle'. See also *cycle*.

animation rostrum (stand) The machine on which all flat animation (cartoons and diagrams) is shot. The term is normally used to include the camera, the camera carriage, the vertical columns, the table and compound, the pantograph, lighting and, where applicable, the floating pegs and aerial image equipment.

animator The artist in an animation studio who prepares the pencil sketches of the key positions of a character's action.

answer print The first sync print of a completed film, after the viewing of which corrections can be made in the grading and colour balance of individual scenes. Second answer print is a second copy if first corrections are inadequate.

aperture (lens aperture) The iris diaphragm in a lens, usually measured in *f* numbers.

aspect ratio Refers to the ratio of width to height of a film frame. Most popular for theatrical projection today is 1:1.85.

audio-visual Equipment, productions and presentations combining sound and picture.

author, script author The author is the writer of the original work on which a film is based, not necessarily a film script but perhaps a novel or play. The script author, more usually called the script writer, is the writer of the film script.

auto-focus Equipment fitted to most animation stands that automatically focuses the camera, whatever field size is being shot. Similar equipment is fitted to two-lens aerial image projection equipment.

back projection The projection of a motion picture, still slide or TV image onto the rear of a translucent screen, to be viewed from its front surface. In animation the technique is used occasionally, projecting onto a very finely ground glass screen and using double-exposure.

background In animation the background is usually painted on paper by an artist specializing in this work. It could also be a three-dimensional model.

background layout A pencil drawing of the background made by the layout artist, from which the background artist makes a finished painting.

bar sheets Sheets that carry groups of horizontal lines on which are vertical marks representing frames. Each word of dialogue, sound effects and notes of music are marked on these sheets so that the frame numbers on which any start or finish of action occurs are accurately recorded. Only when such sheets have been prepared from generally pre-recorded sound tracks can the animator start work and fit lip movements, etc. to the soundtracks.

blank, blank cel If a character on a certain cel level goes out of picture in the middle of a scene a blank cel is substituted on this cel level. If this were not done there would be small colour changes in the background and all lower cel levels in the second part of the scene.

blue screen A method of making mattes in colour cinematography. The subject is shot against a uniformly lit blue background. The laboratory then separates the positive and negative mattes of the image. The mattes can then be used in combination with any filmed background image. Caution must be exercised to avoid the use of blue in the foreground image which results in the phenomenon of 'blue spill'.

budget A document listing all the estimated costs of a film. More generally the amount of money allocated to a particular film production.

calibration The way in which a measuring instrument is graduated. For instance, the controls of the compound are calibrated in hundreds of an inch.

camera Normally a motion-picture camera but can, of course, be a still camera. The animation stand is often loosely called the 'camera' and the room in which it is erected the 'camera room'.

camera animation A motion-picture camera having special adaptations making it suitable for animation photography. These adaptations most often include a single- and continuous-exposure control, forward and reverse directional movement of film, a shuttle movement and registration system, a variable opening shutter with fade scales, and a rackover or reflex viewfinder to permit direct viewing of the animation artwork.

cameraless animation Animation performed by painting directly onto 35 mm or 16 mm blank film. This technique was used by the Australian artist Len Lye and has also been much used by Norman McLaren of the National Film Board of Canada. Arnaldo Ginna pioneered the technique as far back as 1910.

caption Artwork, lettering, photographs or diagrams used for programmes. A written or spoken title identifying or explaining the content of a pictorial image.

cel animation Animation by means of substituting cels, one for each frame or for every two frames, as opposed to mechanical movements performed on the animation stand or hand animation of cutouts and flat models under the camera.

cel level The position of a given cell when several cels are placed on top of each other.

cel punch A punch designed to punch registration holes in cels or paper for backgrounds. While normally only three holes are punched in any one operation, it is also fitted with a pair of pegs which permits the punching of alternate round and rectangular holes throughout the length of a long cel.

cel, standard A transparent plastic sheet whose dimensions are slightly larger than a 12 field; the type most commonly used in animation artwork.

character animation The art of making an animated figure move like a unique individual; sometimes described as acting through drawings. The animator must understand how the character's personality and body structure will be reflected in its movements.

character model A sheet of drawings defining the proportions, shape, clothing etc. of a character for the guidance of animators.

checker An individual who checks the numbering of cels, backgrounds and camera movements against the dope sheet before they reach final photography.

clay animation Animation of three-dimensional puppets or other objects made of plasticine or other malleable material.

close-up A photograph of a detail within a larger picture on the animation stand.

collage A composition of flat objects, such as newspapers, cloth, cardboard, etc. pasted together on a surface and often combined with related lines and colour for artistic effect.

colour key Cel showing names or numbers of colours of an animated character.

colour test Footage of a film that has been timed and which is used as a check to make sure that colours, characters and backgrounds do not clash in the finished film. See *timing*.

compound The mechanism whereby the animation table, the floating pegs and one of the lenses in the two-lens aerial image projection system can be moved in any direction. It consists of two sets of rails or slides, normally but not always at right angles, the second set being carried by the first. The table, floating pegs or lens slides on the second set, so that by making movements on both sets it can be moved in any direction.

computer animation Animation produced by a program in a computer. This is quite different to the use of a computer 'for' animation, where the computer calculates and perhaps controls the mechanical movements of the animation stand. See pages 117–123.

contact printing A method of printing in which the raw stock is held in direct contact with the film bearing the image to be copied.

co-production The production of a film by two or more separate companies, often from different countries.

creeping titles Long titles which pan slowly northwards on the screen, the top lines disappearing as more lines appear at the bottom. Often called 'rolling titles' (American term: crawling titles).

CRI colour reversal internegative, i.e. a duplicate negative.

cutback If a mistake is made in photography a card is shot with instructions to the cutter to 'cut back *X* frames'. The cameraman then reshoots the section correctly, and the incorrect section is cut out of the print before projection.

cutout A small piece of cutout artwork which can be laid on a background under the animation camera and, if required, can be moved frame by frame by hand. Arrows and labels are typical examples of cutouts.

cycle A set of artwork for an action which repeats itself after a certain number of frames. Thus the animation of a walking character may consist of, say, 12 cels. After exposing cels 1 to 12, the cycle is started again with cel 1.

dialogue The speech of actors or characters in a film, as opposed to that of a commentator.

director The person who directs the middle process of making a live-action film, i.e. who directs the actors and camera as opposed to the script writing or editing. In animation, the director will supervise the animators, approve the music, direct the voices, etc.

dissolve, mix The dissolve of one scene into the next in a film, made by superimposing a fade-out of the first scene onto a fade-in of the second.

documentary film A factual film. John Grierson, who invented the term, defined it as 'creative interpretation (on film) of reality'.

dope sheet The correct term is *camera exposure sheet*, or *exposure sheet*, but the term dope sheet is nearly always used. It is a sheet consisting of a large number of lines and a number of columns. Each line represents one frame. There will be columns for each of the cel levels, for background, for peg-bar movements, for pan and zoom movements, and one column for special camera instructions such as fades and mixes. There will be columns for frame numbers and usually two broad columns to the left, one for action (picture) and one for sound, in which the film director writes his instructions to the animator.

double-framing The shooting of two frames of each cel under the animation camera instead of one. This means that only half the number of cels have to be made and the results on the screen are quite satisfactory if the movement is not too rapid. The American term is 'shoot in twos'.

drawn-on-film An animation technique in which the image is drawn, painted or scratched directly onto the film stock.

dry brush A form of rendering in which a brush is dipped into paint or ink and dragged nearly dry before rendering.

dubbing The mixing of speech tracks, sound effects tracks and music tracks onto a single track in the dubbing theatre. The term is also used for making a foreign language speech track to replace that in the original language.

DX double exposure, i.e. the exposure of two or more images on film without the use of mattes.

editor, film editor The person who performs the editing of a film. He or she will be assisted by an assistant editor, and there may be a dubbing editor (also 'sound editor') to take care of the soundtracks.

effects animation The animation of non-character movements such as rain, smoke, lightning, water, etc.

effects track A soundtrack carrying sound effects only. There are generally two or three such tracks for a film, all of which are mixed together with speech and music tracks for the finished film.

fade A camera effect in which the image either gradually appears from black (fade-in) or disappears to black (fade-out).

fairings The acceleration from rest at the start of a movement (initial fairings), the deceleration to rest at the end of a movement (final fairings), or the acceleration or deceleration when a movement changes from one speed to another (middle fairings). Terms such as ease-in, ease-out, buffer, cushion and taper are often used for fairings, but while they are quite suitable for initial and final fairings, the terms are not very acceptable for middle fairings.

field chart, field guide A sheet of plastic, punched for pegs, on which is engraved the exact area of the maximum area that can be used on a certain set-up (e.g. 12 field, 15 field, etc). This area is subdivided by horizontal and vertical lines. The use of a field chart (American term: 'field guide') permits the animator to give instructions to the camera operator, indicating numerically the exact size and position of the field in any scene.

field key A drawing provided by the animator to give the camera operator a visual check on the area or areas to be photographed in a scene.

field size A measure of the size of the area being photographed on the animation stand at a given moment. Field sizes range from a 1 field to a 12 field, and are of a ratio of 30:50.

filmstrip Compilation of individual still photographs, illustrations on a single length of film for projection purposes.

flip An effect whereby a still scene appears to revolve on its vertical axis and may introduce a new scene or a new title with each revolution.

flow chart A graphical representation of the possible alternative routes available in a programme.

focal length For any lens or combination of lens elements such as a photographic objective, the focal length is the distance from either principal focus to the corresponding principal point. The principal points are two positions on the optical axis. separated from each other by a distance variable with the characteristics of the individual lens or combination of lenses. Under ordinary conditions these coincide with the conjugate points for unit magnification, in this context being known as nodal points.

focus The point where parallel rays of light refracted by a given lens appear to meet.

follow focus cam A mechanical device used on an animation stand, with a linkage between the column and the lens of the camera, which keeps the lens focused on the surface of the compound tabletop regardless of the height of the camera above the subject.

footage A given length of motion-picture film.

frame An individual image on a piece of film, representing 1/24th of a second on the screen.

front projection Projection from the front onto a reflecting screen as opposed to projection from the back onto a translucent screen. In particular, in a film studio, projection via a beam-splitter onto a beaded screen which reflects all light back in the same direction from which it came, which is a process now largely replacing back projection.

full animation Animation where a new drawing is done for every frame or every other frame to produce maximum smoothness of action.

graphics Artwork, captions, lettering, photographs, etc. used in programmes.

graticule The standard field key, marked in quarter fields, for working out offset field centres. There are two sizes, 5½ field and 7½ field.

guide lines Lightly drawn pencil lines used to control and direct the movements of animated cutouts and objects.

held cel, hold cel A cel that is held stationary for a number of frames while cels on other levels continue to be animated.

hold The holding stationary of all art work being photographed during an otherwise animated sequence.

holography The recording and representation of three-dimensional objects by interference patterns formed between a beam of coherent light and its reflection from the subject.

inbetween The paper drawing of a figure that lies in sequence between two key positions drawn by an animator.

inbetween breakdown Instructions from the animator to the inbetweener with regard to the treatment of inbetween drawings.

ink and paint The step in cel animation in which the animators' drawings are transferred to cels for photographing. The drawings are first inked by tracing them onto the front of the cels with a pen or fine brush, or transferred by xerographic process. The backs of the cels are then painted, usually with special acrylic paints.

inking The process of tracing the animation drawings on cels in ink.

insert Usually a scene photographed separately and added during editing.

inter-negative A print reproduced from an original film photographed on reversal film stock, and used to print duplicates and release prints.

inter-positive A colour master positive film printed directly from an original colour negative onto intermediate stock.

iris (1) The iris diaphragm in a lens which controls the amount of light which reaches the film or is projected through the lens. (2) *Iris-in*, *Iris-out*: an old-fashioned method of starting or finishing a scene. More and more of the scene is revealed as the iris opens, or is hidden as the iris closes. The iris is not necessarily circular nor centred on the middle of the screen. The effect is an alternative to a fade-in or fade-out.

key animator In studio animation the artist who draws the key or extreme poses: a fully fledged animator.

laboratory, lab The works at which film is developed and printed. Film laboratories have other functions such as optical printing, negative cutting, special optical effects such as travelling mattes, etc.

laser Light Amplification by Stimulated Emission of Radiation. A device for generating a beam of coherent monochromatic radiation.

layout A pencil drawing showing the main features of the background of a scene together with the exact field sizes and positions for the scene. If there are to be match-lines, those parts of the layout must be drawn with great accuracy.

level Up to four and occasionally five cels may be used together on top of the background. The position of a cel among this stack is termed its 'level', the cel next to the background being level 1, the one above this level 2, etc.

light box A drawing desk into which fits a drawing disc. It is lighted from underneath and is used to trace one drawing from the previous one in sequence.

limited animation A film or shot in which there is comparatively little full animation, and therefore much cheaper to make. The technique became popular after the release of Disney's *The Reluctant Dragon*, in which the 'Baby Weems' sequence was basically a photographed storyboard. See also *full animation*.

line film High-contrast film which, after correct development, gives negatives of black and white only. Its blacks are not considered dense enough for many rostrum camera applications.

line test When pencil drawings for an animation scene have been completed they are generally photographed with back lighting if there is more than one level, to ensure that the animation is satisfactory, before they are traced and painted on cels. Only a black and white negative is made of such tests, which are known as line tests. The American term is 'pencil test'.

lip sync The synchronization of lip movements of a character speaking, with the dialogue spoken. In live action the dialogue is normally recorded at the same time as the scene is being filmed, but in animation the dialogue is prerecorded and the animator has to draw the lip movements of his character to match. Animators become very experienced in this work, but are often supplied with a mirror beside their light box so that they can study their own lip movements speaking the appropriate words. P's and B's, being labials with large lip movements, are the most important consonants to get accurately synchronized.

live action A term used to designate a shot of real motion as opposed to an animated shot. The term includes not only shots taken at the normal 24 pictures per second but also slow motion and ultra-fast motion.

long cel A cel used when it is necessary to pan the animation without the cel edge appearing on the screen. The usual length is double that of a normal cel but it can also be three times the normal length, or, if necessary, cut from a roll.

long shot A scene which represents a more or less distant view of the object being photographed or showing the general view of the setting within which the action takes place.

magazine A chamber for holding rolled motion picture film, usually designed with the dual capacity to supply the film for exposure and then reroll it, used with certain types of cameras and printers.

magnetic recording The recording of the sequential sound waves of the audio subject, on a film, tape or wire which has been coated with a magnetic recording medium.

master (1) The original 16 mm reversal film exposed in the camera, after processing. (2) A special positive print made from an original negative for protection or duplication rather than projection. (3) The final version of any programme from which show copies will be made.

match To fit a drawing to another, or to part of the background, so that although it is actually on top it appears to go behind the drawing or part of the background to which it is fitted.

matte A film term meaning 'mask'. Often some part of the field is matted out and some other image inserted there later, a process known as 'image replacement'. See also *travelling matte.* A matte shot in live action can have a live-action foreground shot on a set or location, and a painting, usually painted on glass, in the background.

M & E Music and effects. A soundtrack containing music and effects without speech or dialogue.

mix (1) American: the combining of voices, music, and sound effects into desirable proportions on a single soundtrack. (2) British: see *dissolve.*

mixing (1) *In sound*: mixing or dubbing several soundtracks into one track. (2) *In picture*: mixing or dissolving one scene into the next.

multimedia A presentation using a mixture of media (for example, tape-slide, motion picture, video and live action).

multiplane A very elaborate animation stand having up to seven glass planes on which background and foregrounds are painted, two of which are also designed to accept cels. All planes can be moved E–W, N–S or up–down and the camera can be moved up–down or rotated. Each plane has its own lighting. Shots with a moving camera shot in multiplane appear to have a depth impossible to achieve on a normal animation stand.

music chart A breakdown, frame by frame, of a music track so that an animator can work exactly to the beat.

mute print A print carrying picture only, not sound.

negative perforation Used for 35 mm film for making original negatives or registration duplicates. Has a flat top and bottom which accurately fits the locating pins on a registration film movement.

north/south pegs A special attachment which can be fixed in place of the bottom peg bar under the camera. By means of a screw the cameraman can displace the pegs an inch or so either way in a north or south direction.

object animation The movement or manipulation of three-dimensional objects photographed by single-frame exposure.

ones, twos, or threes The number of frames each drawing is held during filming. The smoothest animation is done on ones and twos, which means 24 or 12 drawings are used per second of screen time. Animating on threes will work for some movements but if the drawings are held for more than three frames the movements will appear jerky or stiff.

optical print A print made on an optical printer as opposed to a contact printer.

overlay A foreground piece in an animation scene, generally but not always painted on a cel, which is registered on an upper level and so permits objects registered on a lower level to pass behind it without resorting to match-lines.

painter A person who specializes in painting the colours on cels.

pan, animation The horizontal or vertical movement of the animation compound, bearing the animation artwork, while being photographed.

pan shot Derived from 'panoramic'. A shot which encompasses a wider area than can be viewed by the camera at one time, and which will be scanned by the camera by means of panning.

panning board A board with sliding peg bars, which an animator can put over a light box when working out peg movements. There are panning boards with 5½ peg bars and 7½ peg bars.

peg animation A term used for that type of animation in which the illusion of motion is produced by substituting one piece of artwork for another every frame or every second frame, the artwork being registered on pegs.

pegs Small metal or plastic projections affixed to all surfaces that will support the artwork during production. They correspond to holes punched in drawing paper, cels and background artwork and are used to maintain registration through all stages of the production.

pegs centre The position when the middle peg of a set is in line with the vertical centre of the standard field. Peg settings are given in inches left or right of this centre. Panning peg moves are given in decimals of an inch: 'pegs left' or 'pegs right'.

pencil test The process of photographing and projecting the pencilled animation drawings to determine the smoothness of the animation before proceeding to the inking and painting of cels.

photography area The portion of the surface of the compound tabletop of an animation stand which is within the photographic scope of the animation camera.

pin-screen animation An ingenious system devised by Alexeieff and Parker, consisting of a plate containing a very large number of closely spaced holes, each of which contains a pin, and which is lighted from two lamps at the top. If the pins are pulled fairly far out they each cast two shadows and form a dark area. Every pin thus has two shadows, 60° inclined. Given that the pins are disposed in triangles (equilateral), every shadow is inclined directly on the next pin, until it is long enough to cover it and to produce the dark surface. If the pins are pushed flush with the surface of the plate, a light area results. Images or pictures can therefore be made by moving the pins, and such images can be animated. The results look rather like animated engraving. Alexeieff's most famous film using this technique was *Night on a Bare Mountain*, set to Mussorgsky's music.

pixilation Animation using living actors instead of artwork or puppets. A series of posed positions photographed frame by frame can result in astonishing and effective shots. Mostly used for comic effects, but such films can also have serious themes.

platen, platen glass A piece of glass mounted in a metal frame which can be lowered onto artwork being photographed on an animation stand, to keep the artwork flat while being photographed.

post-syncing The recording, mostly of dialogue, in sync with lip movements after the picture has been shot. Animated films normally have speech and music prerecorded.

prerecording The recording of speech and music before the picture has been shot. This is the normal procedure with cartoon films, since it is obviously much easier for an animator to make animation exactly fit prerecorded speech and music than to make such speech and music fit an animated scene later.

production manager A person who organizes the production of a film, booking studios, arranging for the building of sets, engaging staff and artists, helping to schedule the production of the film and so on. He or she will usually be assisted by a unit manager and, for location work, by a location manager.

punch Mechanical device used to punch registration holes in cels and paper.

puppet animation The animation of puppets. Strictly speaking, this is only the use of puppets shot one frame at a time using either articulated puppets or substituting a whole puppet in a different attitude each frame (the George Pal system). String puppets (marionettes) or glove puppets shot at live-action speed is not puppet animation.

rear projection Motion-picture images projected onto the desired screen surface, usually translucent glass, from behind the surface instead of from the front.

register pegs Metal dowels on the table of the animation stand on which the artwork is registered, or similar pegs on the light box which enables the animator or inbetweener to trace, with small alterations, from the previous drawing of a character.

registration The correct relationship between the positions of the animation artwork and the camera.

repeat (cycle) A short series of drawings or cels which may be photographed over and over to create the illusion of continuing, repeated action.

reversal film Film which, after exposure, is processed to produce a positive image rather than the customary negative image. If exposure is made by printing from a negative, a negative image is produced. Reversal films may be black and white, or colour, and either sound or picture, or both; usually 16 mm films.

reverse action The result of a scene shot backwards. Sometimes the reverse action is obvious and is used as a comic effect. In animation, *scratchback* effects are normally shot with the camera running backwards.

rostrum camera A fixed film or TV camera mounted vertically for shooting graphics and animation.

rotoscope A device which enables a piece of developed film loaded into the animation camera to be projected down onto the table of the animation stand frame by frame. Among many uses, the action can be traced or copied.

rough cut A stage between the assembly of shots into a cutting copy and the fine cutting that is done later.

rushes The first print of a day's shooting returned by the laboratory, usually overnight. The American term is 'dailies'.

scale A diagram made by an animator in the corner of a key drawing to indicate the number and spacing of intermediate drawings needed between this key and the next.

scene A scene or script scene is a numbered part of a film script which may be broken down into parts in long-shot, medium-shot, close-up, etc. by the director when shooting. A master scene is a fairly long length of the script, all under one number, which the director will certainly break down later. He or she may, however, take the whole of a master scene first, then shoot close-ups of the various characters to cut in with this later. In animation the basic unit of continuous action, usually shot on one background, from which a film is built up.

scratchback The technique in which parts of the artwork, painted on cel, are removed frame by frame under the camera while the latter runs backwards. The effect is that the artwork appears to grow on the screen.

script The detailed scene-by-scene instructions for a film or television production, including description of setting and action with dialogue and camera directions. When the script also has full details of visuals it is termed a 'storyboard'.

sequence A considerable portion of a film script or the corresponding portion of the film itself when edited, including perhaps many script scenes. Roughly equivalent to a scene in a live theatre play. An animated sequence is a section of animation inserted into a live-action film.

short The term usually refers to the cartoons made in Hollywood during the 1930s, 1940s and 1950s, which ran between 6 and 7 minutes. Today, shorts range from 1½ to over 20 minutes in length and cover a variety of styles and subjects.

silent speed The minimum rate of film projection at which the projected images can be fused by the persistence of vision into a continuous image, considered to be 16 frames per second.

single-frame exposure The exposure of one frame of motion picture film at a time, in the manner of still photography. Commonly used in animation because of the need to make changes of cels or in the relationship of the compound to the camera.

slit scan The technique which achieves a specific effect of distortion by photographing a subject in motion one portion at a time. The effect is the ability to add perspective (to a two-dimensional image) that changes naturally as the point of view moves. While the camera shutter holds open, a slit 'scans' across the image or the film, recording only a carefully controlled portion of the image for each position of the camera move. This is repeated (with subtle sequential variations) for each frame of film. The shape of the slit is limitless, and can be changed or animated, frame by frame, during the shot.

slow in/slow out Refers to the fact that panning and trucking moves usually begin slowly, gradually attain their full speed and then slow to a stop, to avoid a sense of jerkiness in the movement.

sound effects Noises and sounds for a film other than speech or music.

sound speed The standard rate of projection for 16 mm sound motion pictures is 24 frames per second or 36 feet per minute. The comparable rate for 35 mm sound film is 24 frames per second (90 feet per minute).

special effects animation Animation of visual effects such as smoke or water. Also animation of models by stop-motion in live action.

speed lines Lines drawn in the direction taken by an animated subject in order to emphasize its speed.

split screen Two or more independent pictures composed on one frame and divided by a sharp line.

sprocket hole A perforation on the edge of a film.

squash and stretch An element of character animation which involves the exaggeration of the normal tendency of an object in motion to undergo a degree of distortion, lengthening as it travels and compressing as it stops.

stereoscopic animation Animated films which achieve a three-dimensional effect by presenting multiple images to the spectator but require the viewer to wear spectacles in order to resolve the images. Vectographic film, which produced a stereoscopic effect without the need of viewing glasses, proved too thick to be feasible in commercial processing.

storyboard A form of shooting script, common for animated films for many years and now usually used for commercials, even live-action ones. It consists of a series of sketches showing key positions for every scene, with dialogue and descriptive notes below. Still used in animation.

stroboscope Pulsed light source used to measure speed of rotation or frequency of repetition or to freeze the motion of objects.

studio animation Animation characterized by the fact that various aspects of production are done by different people and, in the case of larger studios, different departments. The finished product is the result of a coordinated group effort.

superimposition The result on the screen of two images appearing on top of each other. This is usually done by double-exposure, but double-exposure does not necessarily result in superimposition, as, for example, in travelling matte shots.

television animation Animated films intended for broadcast distribution differ from the traditional form in that their subject must be more carefully centred to preclude any amputation of key images by 'television cutoff'.

television cutoff The area surrounding the edges of each frame of film which is masked off by television projection.

test In live action a set of shots of an actor made to help decide whether he or she is suitable for the part. The term 'test' is also used for a shot made to test the correct functioning of a camera, a test shot to try out some special device, etc. See also *line test*.

top peg A procedure used in cases where one piece of artwork must traverse another. The second piece of artwork is attached to a peg along the top edge, thereby allowing the two pieces of artwork to pair accurately with each other (e.g. moiré, slitscan effects).

tracing The process of tracing the pencil drawings of an animator or inbetweener onto the front of cels. Sometimes called 'inking'.

travelling matte A black silhouette on a piece of high-contrast film which moves its position and possibly its shape from frame to frame. When loaded in a camera emulsion to emulsion with undeveloped film stock, it masks off a changing area of the scene being photographed. Then the character or object from which the travelling matte was made, or a film shot of the same, can be photographed onto the same piece of film emulsion to emulsion with a complementary travelling matte. The result will be a composite shot of the matted object in a separately photographed background. In live-action the composite shot is made on an optical printer. In animation it is usually made in the animation camera. Travelling matte is simply a term meaning 'moving mask'.

travelling peg bars The peg bars recessed into the animation table which can be moved mechanically in an E–W direction.

treatment An initial description of a film in narrative form, written before the film script is prepared.

twos The exposure of two frames of motion picture film to a given image.

video animation stand A unit designed to photograph animation on video tape. The recorder moves the tape in 1/24th of a second increments that correspond to frames of film. Instant replay capacity makes it useful for line tests.

voiceover Narration recorded without the speaker being seen.

voice track A recording track for narration or dialogue as opposed to music and effects.

wide screen Any screen which uses a high aspect-ratio, such as Cinemascope, Vistavision or cropped 35 mm using a wider aspect-ratio than 1.33:1.

work print (cutting copy) The print of assembled rushes (dailies) on which the editor works until he or she has the final version of the film. After this, the soundtracks are dubbed and the picture negative is cut to match.

xerography The copying of documents onto plain paper or cel using a photoelectric effect, normally a plate or cylinder of selenium. The process is also used for copying the pencil drawings of the animator onto cels.

zoom In animation, the movement of the camera on the animation stand towards or away from the artwork.

Computer

algorithm A set of well-defined rules for the solution of a problem. Algorithms are implemented on a computer by a stored sequence of instructions. The method for computing the square root of a number is an algorithm.

aliasing Undesirable visual effect in computer generated images, caused by inadequate sampling to completely define the form; most commonly a stepped edge or 'staircase' along the object boundary.

analog The characteristic of varying continually along a scale, as opposed to increasing or decreasing in fixed increments or steps. Voltage, pressure, speed etc. are often measured in analog terms.

analog computer Where the result is shown by analogy, e.g. increasing stress may be shown by increasing voltage, as against a digital computer, which shows the results as numbers.

array Data structure which may store the location of points by coordinates or some other group of related variables; a matrix. For example, a three-dimensional array can store x, y, z coordinates.

BASIC Beginner's All-purpose Symbolic Instruction Code. A high-level programming language, resembling English. One of the easier programming languages for use by non-experts.

batch processing A technique in which items to be processed in identical or similar ways are collected into groups (batched) to permit convenient and efficient processing.

binary code A code that makes use of only two distinct characters, usually 1 and 0, on or off.

bit A binary digit having the value of either 1 or 0. The basic unit of storage in a digital computer.

Boolean Pertaining to the symbolic logic system developed by British mathematician George Boole, containing the notions of 'and', 'or', 'not' and 'exclusive or'.

buffer An intermediate storage area or device that holds data temporarily that are being transferred between elements of a computer system. It is usually an area of memory but it may be a disk or tape.

bug A software error which causes an unintended result.

byte A set of continuous binary bits, usually eight, which represent a character, symbol or operation.

CAD Computer Aided Design. The use of the computer to generate or originate graphic material.

cathode ray tube An evacuated glass tube in which a beam of electrons is emitted and focused onto a phosphor-coated surface to create images, i.e. a television tube.

chip A small, fingernail-sized integrated circuit (IC) package usually containing thousands of logic elements; a 4K device has about 4000 circuits; a 6K device about 64 000.

COBOL Common Business Orientated Language. A high-level programming language designed primarily for business or commercial use.

code A symbol or set of symbols used to represent data or instructions to a computer. Also the actual writing of source instructions to be translated to machine instructions.

computer A device capable of producing useful information or functions by accepting data and performing prescribed operations. Various types of computers are calculators, digital computers and analog computers.

computer animation The technique of using computers to generate moving pictures. Some systems can achieve this in real-time (25 frames per second – or in the USA 30 fps), but the majority of animation is created one frame at a time and then edited into a continuous sequence. Very sophisticated programs are required to perform the tasks of movement, fairing, perspective, hidden-surface removal, colouring, shading and illumination, and as the trend increases towards more realistic images, faster computers are needed to process the millions of computations required for each frame.

computer graphics Charts, diagrams, drawings and other pictorial representations that are computer generated.

computer language In order to communicate with a computer it is necessary to know a specific vocabulary and the way of using it. This would generally take the form of written instructions which can be interpreted and obeyed by the machine.

computer-generated animation Animated images are created inside the computer and exist first as digital coding rather than as drawn images on paper. The coding is then used to control an output device like a film recorder.

coordinate One or several numerical qualities which, between them, define the position of a point on a plan or in space.

CRT Cathode ray tube.

data bank The mass storage of a large amount of information indexed in a manner that facilitates selective retrieval.

data base Data items that must be previously stored in order to meet specified information processing and retrieval needs. The term generally applies to an integrated file of data. A data bank is a data base.

data structure A computer file organization designed so that relationships between data elements are preserved.

density (1) The relative saturation of an area with a quality, e.g. chroma or blackness (value). (2) The amount of information which can be stored on a medium in a given area, e.g. on magnetic tape a common density is 800 bytes or characters per inch.

digitizer A graphic input peripheral which can be used to scan an existing analog graphic image, capturing x, y coordinates in digital form at desired intervals.

display buffer Storage or memory that holds all data required to generate an image.

encoder A device which reads out the angular position of a shaft in a digital form. Encoders can be used to control the position of a picture element on a CRT screen, including its rotation about any desired axis.

floppy disc A thin, flexible circular film with a magnetic surface enclosed within a heavy paper cover, capable of storing digitized information. An inexpensive form of memory, easily inserted and removed from a computer system.

FORTRAN FORmula TRANslator. A high-level language oriented towards mathematical operations.

frame buffer Holding area in computer memory for x, y pixel data to be displayed in one raster scan of a CRT. The depth of the frame buffer (number of bits per pixel) determines the number of colours or intensities which can be displayed.

graphic display device A display terminal or monitor used to display data in a graphic form. The most common types are direct view storage tubes (DVST), raster refresh devices and vector (sometimes called calligraphic or stroke-writing) refresh devices.

graphics packages Graphics packages such as GINO, PICASO, MOVIE, BYU, DISPLA, etc. have been designed to ease certain problems in computer graphics. The user can interact in real-time with the package or prepare a program or command file which can be processed to produce the desired images.

hardware The actual equipment which makes up a computer system, CPU, integrated circuits, etc. as opposed to the program or instructions for the system (software).

hidden line removal Software written to eliminate lines that would normally be hidden from view in the presentation of a solid object.

hologram A three-dimensional image created by illuminating a scene with coherent light derived from a laser and allowing the reflected light to fall upon a photographic plate which is also illuminated by the original light source.

hybrid computer A combination of two distinctly different types of computer which takes advantage of the most useful and applicable features of each. A digital computer operates using numbers (i.e. it counts and tabulates) and has a large informational storage capacity, whereas the analog computer uses the relationships between electrical voltages to formulate results and also has a much greater flexibility in reacting to constantly changing situations.

input The raw data, text or commands inserted into a computer.

input devices Light pens, graphics tablets, keyboards, touch-sensitive screens, joysticks or any device used to give a computer alphanumeric or graphical information.

interactive Immediate or quick response to input. In interactive processing an image can be modified or edited and the changes seen immediately, in contrast to batch processing, in which the user must wait for results.

joystick A lever that can be moved in an x, y direction to control the movement of one or more display elements.

key frame animation The animator 'draws' directly onto the CRT display and produces a basic picture or cel. A number of these drawings can then be superimposed on one another to form a composite cel or key frame. Many of these key frames can be made up and stored in the computer to be called up and used as required. The action of the film can be created by stringing together the series of key frames, and introducing the desired movements between one frame and the next. Each key frame can be used over and over again by simply calling it repeatedly from the computer score.

language A set of representations, conventions and rules used to convey information.

light pen A highly sensitive photo-electric device used with a visual display unit. The operator can pass the pen over the surface of the screen to detect images displayed on the screen, or to change or modify images which have been displayed.

logarithm Of a number, the power of a fixed number (called the base, usually 10), that equals the number.

logic The use of program operators to perform logical operations on binary numbers.

machine language Coded language used directly by a computer in which all commands are expressed as a series of 1's and 0's. Thus a machine language program is a set of instructions which a computer can recognize directly without intermediate interpretation.

mainframe Central processing unit of a computer.

memory The area in a computer into which data can be entered and stored for subsequent retrieval.

menu A list on the face of the CRT used to display options or choices.

microcomputer A small computer containing a microprocessor, input and display devices, and memory, all in one box. It may or may not interface to a host computer and/or peripheral devices. Sometimes referred to as a desktop computer.

mode A particular method of operation for a hardware or software device. For example, a particular unit might be capable of operating in, say, binary mode or character mode.

modelling An abstract mathematical description of a graphic object that a computer can translate into a graphic display in 2D or 3D.

mouse A small device that is moved over a surface and which, by its position and motion, provides coordinate input to the display device.

NTSC National Television Standards Committee; the American colour TV coding system, 525 lines, 60 fields, 30 frames per second.

off line Isolated, not connected to the line, central computer system, etc.

on line Equipment in direct communication with the central processor of a computer system, as opposed to off-line devices.

output The material produced by a device such as a computer, e.g. printout, magnetic tape, etc.

pad In computer graphics, a data tablet.

paint system A colour graphics microcomputer, usually including a digitizing tablet, that allows an artist to draw directly into the computer frame buffer, the results of which are displayed on a CRT and able to be transferred to video, film, recorder or other devices.

PAL (1) Programmable Array Logic. (2) Phase Alternation Line; the European colour television coding system, 625 line, 50 fields, 25 frames per second.

palette Interactive computer graphics systems which enable the user to handle colour in a real-time mode should allow the user to define and mix new colours when required. This can be achieved by displaying upon the colour terminal a palette (selection) of colours which can form the basis of new colours when two or more are selected and mixed together. Paint Box systems permit this colour handling and the palette can be stored for future use.

peripherals External items added to the computer system, e.g. floppy disc drives, printers, visual display units, etc.

pixel Picture element. The smallest visual unit on the screen which can be stored, displayed or modified. Limited by the resolution of the display device. For example, a screen with a resolution of 512 × 512 lines will contain 262 144 pixels.

plotter A graphic hard-copy output device which can use any of a number of technologies to form an image. The plotter uses a point contact or projection onto a medium, such as paper, to draw a picture or character under computer control. The plotting head can move over the entire surface of the drawing medium or be restricted to a linear area. The line between printers and plotters is very narrow. Some plotters can create characters and printers can draw pictures. Pen, electrostatic, photo, ink-jet and laser plotters are some examples.

polygon A two-dimensional geometrical figure bounded by straight lines.

printout The printed output of a computer, i.e. a paper record of the computer's computations and processing.

RAM Random access memory. Data storage in which records of individual bytes are accessible independent of their location in relation to the previous record or byte accessed.

raster A display device that stores and displays data as a sequential series of horizontal rows of picture elements (pixels).

ray tracing A technique for producing computer-generated shaded images.

real-time Performance of a computerized operation that gives the impression of instantaneous response.

resolution (1) Number of pixels per unit of area. A display with a finer grid contains more pixels and thus has a higher resolution, capable of reproducing more detail in an image. (2) The smallest distance between two elements which can be perceived as distinct by the viewer.

saturation A subjective term usually referring to the difference of a hue from a grey of the same value, i.e. chromaticity. In a subtractive system, adding the complement will make the colour darker. In an additive system, adding the complement will make the colour lighter. Purity is an objective term which denotes a measurement that can be visualized on a chromacity chart as a position between the equal energy mixture of all colours (white) to the dominant wavelength of the colour.

scanning The process by which an area is systematically explored line by line, particularly in television, where an electron beam is used.

SECAM Séquentiel Couleur à Mémoire (sequential colour with memory). French television system.

SIGGRAPH Special Interest Graphics of the American Association for Computing Machinery. Holds an annual convention for computer scientists and artists interested in computer graphics.

software A program or collection of programs (including assemblers, compilers, utility routines and operating systems) which controls the operation of the computer.

solid state A broad family of electronic devices made solely of solid materials which control current without the use of moving parts, heated filaments or vacuum gaps.

stylus In computer graphics, a pencil-like device for the input of information from a data tablet or for selection from a menu.

subroutine Part of a program which performs a logical section of the overall function of the program. Subroutines may be written for a specific program or in a general form to carry out operations common to several programs.

tablet A tablet, or digitizer, is a device which can convert a stylus position into Cartesian coordinates and, when connected to a graphic display screen, can control the real-time positioning of a cursor.

terminal A point in the system or communication network at which data can be entered, transferred or displayed.

texture mapping A technique for making computer-generated images more realistic.

three-dimensional modelling Geometrical descriptions of an object using polygons or solids in three dimensions (x, y, z coordinates) for the purpose of creating the illusion of height, width and depth.

time encoding A simple method of achieving computer animated images by writing successive images from a frame store to a video recorder. The video tape is formatted with a time code so that the video recorder can rewind over past frames and search forward to the next free position. When the tape is replayed the images are animated in real-time.

turnkey system Computer system containing all the hardware and software needed to perform a given application.

U-matic Video cassette system using ¾-inch tape (Trade name).

user friendly The application of human factors and knowledge to the design of equipment, eliminating the jargon and complexity associated with the early development of machines. Programs in which the user is provided with instructions or prompting for performing most operations.

vector (1) A representation of magnitude and direction. (2) A line connecting defined points on a CRT. (3) A pointer used to allow access to relocatable data or program.

vector graphics The area of computer graphics dedicated to processing and displaying images composed of lines (vectors) as opposed to raster graphics, which creates pictures on a pixel basis.

video cassette A cassette containing video recording tape with separate supply and take-up spools.

videodisc A video and audio recording on a flat rotating medium, usually using digital storage, by optical (reflected laser) or mechanical (capacitive) means.

videotape Magnetic tape specifically designed for use as a video recording medium.

wire frame A three-dimensional image displayed as a series of line segments outlining its surface. Display of an object as a framework consisting of straight-line segments.

workstation A configuration of computer equipment designed to be used by one person at a time. A workstation may have a terminal connected to a larger computer or may be a 'stand-alone' with local processing capability. It usually consists of an input device, e.g. keyboard, digitizer, etc., a display device, memory, and an output device such as a printer or plotter.

X-Y digitizer A device which can convert a physical quantity, i.e. an analog measurement, into coded character form for visual display.

X-Y plotter Also known as a data plotter. A device designed to give a visual display, usually in the form of a graph on paper, by plotting the course of coordinates.

X-Y-Z axis Terms used to describe position or form of an object on a computer display. X is the horizontal plane, Y is the vertical plane and Z is the depth. For example, we can rotate an object around its X axis.

Bibliography

Adamson, Joe, *Tex Avery: King of Cartoons*, New York: Popular Library (1975); New York: Da Capo, paperback (1985)

Anderson, Yvonne, *Make Your Own Animated Movies: Yellow Ball Workshop Film Techniques*, Boston: Little Brown (1970)

Arnheim, Rudolph, *Art and Visual Perception: the Psychology of the Creative Eye*, Berkeley: University of California Press (1965); Boston: Little Brown (1970)

Asenin, Serge, *The Animated Picture*, Moscow (1974)

Bakedano, José. *Norman McLaren: Obra Completa* (in Spanish), Bilbao: Museo de Bellas Artes (1988)

Bain, David and Bruce Harris, *Mickey Mouse: Fifty Happy Years*, New York: Crown (1978)

Barton, C.H., *How to Animate Cut Outs*, London: Focal Press (1955)

Becker, Stephen, *Comic Art in America*, New York: Simon & Schuster (1959)

Blair, Preston, *Animation: Learn How to Draw Animated Cartoons*, Laguna Beach: Foster (1949)

Burder, John, *The Technique of Editing 16 mm Films*, London: Focal Press (1988)

Cabarga, Leslie, *The Fleischer Story*, New York: Crown (1966); Da Capo, paperback (1988)

Canemaker, John, *The Animated Raggedy Ann and Andy*, New York: Bobbs Merill (1977)

Collins, Maynard, *Norman McLaren*, Canadian Film Institute (1976)

Crafton, Donald, *Before Mickey: the Animated Film 1898–1928*, Cambridge, Mass.: MIT Press (1982)

Edera, Bruno, *Full Length Animated Feature Films*, New York: Hastings House (1977)

Feild, Robert D., *The Art of Walt Disney: from Mickey Mouse to the Magic Kingdom*, New York: Abrams (1973)

Graber, Sheila, *Animation is Fun*, Newcastle: Tyneside Cinema (1985)

Gross, Yoram, *The First Animated Step*, Sydney: Martin Educational (1975)

Halas, John, *Design in Motion*, London: Studio Vista (1962)

Halas, John (ed.), *Visual Scripting*. London: Focal Press (1973)

Halas, John, *Computer Animation*, London: Focal Press (1976)

Halas, John, *Film Animation: a Simplified Approach*, Paris: Unesco (1977)

Halas, John, *Graphics in Motion*, Munich: Novum (1982)

Halas, John, *Masters of Animation*, London: BBC Books (1987)

Halas, John and Walter Herdeg, *Film and TV Graphics*, Zurich: Graphis (1967)

Halas, John and Roger Manvell, *The Technique of Film Animation*, London: Focal Press (1968)

Halas, John and Roger Manvell, *Art in Movement: New Directions in Animation*, London: Studio Vista (1970)

Harryhausen, Ray, *Film Fantasy Scrapbook*, New York: Barnes (1972)
Hayward, Stan, *Scriptwriting for Animation*, London: Focal Press (1977)
Heraldson, Donald, *Creators of Life: a History of Animation*, New York: Drake (1975)
Herdeg, Walter, *Film and TV Graphics 2*, New York: Hastings House (1978)
Hollis, Richard and Brian Sibley, *The Disney Studio Story*, London: Octopus (1988)
Holloway, Ronald, *Z is for Zagreb*, New York: Barnes (1972)
Jankel, Annabel and Rocky Morton, *Creative Computer Graphics*, Cambridge: Cambridge University Press (1985)
Laybourne, Kit, *The Animation Book*, New York: Crown (1979)
Levitan, Eli, *Animation Art in the Commercial Film*, New York: Reinhold (1960)
Levitan, Eli, *Handbook of Animation Techniques*, New York: Van Nostrand (1979)
Lutz, Edwin G., *Animated Cartoons. How they are Made. Their Origin and Development*, New York: Gordon (1976)
Madsen, Roy P., *Animated Films: Concepts, Methods, Uses*, New York: Interland (1969)
Maelstaf, R., *The Animated Cartoon in Belgium*, Brussels: MFA (1970)
Manvell, Roger, *The Animated Film*, London: Focal Press (1955)
Manvell, Roger, *Art and Animation*, London: Tantivy Press (1980)
McCay, Winsor, *Little Nemo 1905–1906*, New York: Nostalgia Press (1976)
McLaren, Norman, *Cameraless Animation*, National Film Board of Canda (1958)
McLaren, Norman, *The Drawings if Norman McLaren*, Montreal: Tunda Books (1975)
Muybridge, Eadweard, *Animals in Motion and Human Figures in Motion*, New York: Dover (1955)
Noake, Roger, *Animation: a Guide to Animated Film Techniques*, London: Macdonald-Orbis (1988)
Perisic, Zoran, *The Animation Stand*, London: Focal Press (1976)
Perisic, Zoran, *The Focal Guide to Shooting Animation*, London: Focal Press (1978)
Pilling, Jane (ed.), *That's Not All Folks: A Primer in Cartoonal Knowledge*, London: British Film Institute (1986)
Reiniger, Lotte, *Shadow Theatres and Shadow Films*, New York: Watson Guptill (1970)
Russett, Robert and Cecile Starr, *Experimental Animation: Origin of a New Art*, New York: Da Capo Press, new revised edition (1988)
Salt, Brian G.D., *Movements in Animation*, Oxford: Pergamon Press (1976)
Salt, Brian G.D., *Basic Animation Stand Techniques*, Oxford: Pergamon Press (1977)
Schickel, Richard, *The Disney Version*, London: Michael Joseph (1986)
Stephenson, Ralph, *The Animated Film*, London: Tantivy Press (1973)
Solomon, C. and Stark, R., *The Complete Kodak Animation Book*, Eastman-Kodak (1984)
Thomas, Frank and Ollie Johnston, *Disney Animation: The Illusion of Life*, London: Orbis (1984)
Whitaker, Harold and John Halas, *Timing for Animation*, London: Focal Press (1981)
White, Tony, *The Animator's Workbook*, Oxford: Phaidon (1986)

Index